Computational Intelligence in Healthcare

Computational intelligence (CI) refers to the ability of computers to accomplish tasks that are normally completed by intelligent beings such as humans and animals. Artificial intelligent systems offer great improvement in healthcare systems by providing more intelligent and convenient solutions and services assisted by machine learning, wireless communications, data analytics, cognitive computing, and mobile computing. Modern health treatments are faced with the challenge of acquiring, analysing, and applying the large amount of knowledge necessary to solve complex problems. AI techniques are being effectively used in the field of healthcare systems by extracting the useful information from the vast amounts of data by applying human expertise and CI methods, such as fuzzy models, artificial neural networks, evolutionary algorithms, and probabilistic methods which have recently emerged as promising tools for the development and application of intelligent systems in healthcare practice.

This book starts with the fundamentals of computer intelligence and the techniques and procedures associated with them. Contained in the book are state-of-the-art CI methods and other allied techniques used in healthcare systems as well as advances in different CI methods that confront the problem of effective data analysis and storage faced by healthcare institutions.

The objective of this book is to provide the latest research related to the healthcare sector to researchers and engineers with a platform encompassing state-of-the-art innovations, research and design, and the implementation of methodologies.

Innovations in Intelligent Internet of Everything (IoE)
Series Editor: Fadi Al-Turjman

Computational Intelligence in Healthcare: Applications, Challenges, and Management
Meenu Gupta, Shakeel Ahmed, Rakesh Kumar, and Chadi Altrjman

Blockchain, IOT and AI technologies for Supply Chain Management
Priyanka Chawla, Adarsh Kumar, Anand Nayyar, and Mohd Naved

For more information about the series, please visit: https://www.routledge.com/Innovations-in-Intelligent-Internet-of-Everything-IoE/book-series/IOE

Computational Intelligence in Healthcare

Applications, Challenges, and Management

Edited by
Meenu Gupta
Chandigarh University, India
Shakeel Ahmed
King Faisal University, Saudi Arabia
Rakesh Kumar
Chandigarh University, India
Chadi Altrjman
Near East University, Cyprus

CRC Press
Taylor & Francis Group
Boca Raton London New York

CRC Press is an imprint of the
Taylor & Francis Group, an **informa** business

First edition published 2023
by CRC Press
4 Park Square, Milton Park, Abingdon, Oxon, OX14 4RN

and by CRC Press
6000 Broken Sound Parkway NW, Suite 300, Boca Raton, FL
33487-2742

CRC Press is an imprint of Taylor & Francis LLC

Library of Congress Cataloging-in-Publication Data
Names: Gupta, Meenu, (Computer Science Professor), editor.
Title: Computational intelligence in healthcare : applications, challenges and management / edited By Meenu Gupta, Chandigarh University, India, Shakeel Ahmed, King Faisal University, Saudi Arabia Rakesh Kumar, Chandigarh University, India, Chadi Altrjman, Near East University, Cyprus.
Description: First edition. | Abingdon, Oxon ; Boca Raton, FL : CRC Press, 2023. | Series: Innovations in Intelligent Internet of Everything (IoE) | Includes bibliographical references and index.
Identifiers: LCCN 2022036477 (print) | LCCN 2022036478 (ebook) | ISBN 9781032304939 (hbk) | ISBN 9781032304946 (pbk) | ISBN 9781003305347 (ebk)
Subjects: LCSH: Artificial intelligence—Medical applications. | Medical technology.
Classification: LCC R859.7.A78 C645 2023 (print) | LCC R859.7.A78 (ebook) | DDC 610.285—dc23/eng/20220919
LC record available at https://lccn.loc.gov/2022036477
LC ebook record available at https://lccn.loc.gov/2022036478

ISBN: 978-1-032-30493-9 (hbk)
ISBN: 978-1-032-30494-6 (pbk)
ISBN: 978-1-003-30534-7 (ebk)

DOI: 10.1201/9781003305347

Typeset in Sabon
by Apex CoVantage, LLC

Contents

Editors vii
Contributors ix

1 Introduction to computational intelligence in healthcare: Applications, challenges, and management **1**

CHANDER PRABHA, JASPREET SINGH, SHWETA AGARWAL, AMIT VERMA, AND NEHA SHARMA

2 Role of IoT and machine learning for e-healthcare management **17**

NITIKA PHULL, PARMINDER SINGH, AND KUSRINI

3 Telemedical and remote healthcare monitoring using IoT and machine learning **47**

SHARAD CHAUHAN, KANIKA PAHWA, AND SHAKEEL AHMED

4 Efficient ways for healthcare data management using data science and machine learning **67**

DIMPLE CHEHAL AND PAYAL GULATI

5 A novel scheme to manage the e-healthcare system using cloud computing and the internet of things **81**

NISHA AND MEENU GUPTA

6 Automating remote point-of-care ECG diagnostics via decentralized report routing algorithm **99**

BIDYUT BIKASH BORAH, SATYABRAT MALLA BUJAR BARUAH, DEBARAJ KAKATI, AND SOUMIK ROY

7 **Evaluation of deep image embedders for healthcare informatics improvement using visualized performance metrics** 115

T. O. OLALEYE, A. O. OKEWALE, I. SOLANKE, O. V. ALOMAJA, O. F. ADEBAYO, AND S. M. AKINTUNDE

8 **A comparative analysis for analysing the performance of convolutional neural network versus other machine learning techniques to assess cardiovascular disease** 137

HARSHAVARDHAN TIWARI, AISHWARYA M, HARSHITHA M V, NIDA SHAFIN, SRUSHTI J, AND TUSHITA S

9 **A study of machine learning initiatives in the global healthcare sector using different case studies** 153

SWATI SINGH

10 **Autism spectrum disorder diagnostic system using adaptive neuro fuzzy inference system** 167

JOY KARAN SINGH AND DEEPTI KAKKAR

11 **Detection of diabetic foot ulcer (DFU) with AlexNet and ResNet-101** 181

HASSANA ABUBAKAR, ZUBAIDA SA'ID AMEEN, SINEM ALTURJMAN, AUWALU SALEH MUBARAK, AND FADI AL-TURJMAN

12 **A case study–based analysis on remote medical monitoring with AWS cloud and internet of things (IoT)** 193

GOKUL H, ATHARVA DESHMUKH, SHRADDHA JATHAR, AND AMIT KUMAR TYAGI

Index 213

Editors

Meenu Gupta, PhD, is Associate Professor at the UIE-CSE Department, Chandigarh University, India. She earned a PhD in computer science and engineering with an emphasis on traffic accident severity problems from Ansal University, Gurgaon, India, in 2020. She has more than 13 years of teaching experience. Her areas of research cover the fields of machine learning, intelligent systems, and data mining, with a specific interest in artificial intelligence, smart cities, and others. She has edited two books on healthcare and cancer diseases and authored four engineering books. She works as a reviewer for several journals, including *Soft Computing*, *Big Data*, *Scientific Report*, and others. She is a life member of ISTE and IAENG. She has authored or co-authored more than 20 book chapters and over 60 papers in refereed international journals and conferences. She has filed three Indian patents and was awarded as the best faculty and researcher of the department in 2021.

Shakeel Ahmed, PhD, is Associate Professor at the College of Computer Sciences and Information Technology, King Faisal University, Saudi Arabia. His areas of interest include software verification and validation, mobile ad hoc networks, software engineering, and cloud computing. He is Head of the Cloud Computing Research Group at the College of Computer Sciences and Information Technology, King Faisal University (KFU). He has authored several research papers in indexed and impact factor research journals and IEEE international conferences. He is a PC member for several international conferences and an active reviewer for quartile journals. He earned a PhD in computer science at Indore University, India.

Rakesh Kumar, PhD, is Professor in the Department of Computer Science Engineering at Chandigarh University, Punjab, India. He earned a PhD in computer science and engineering at Punjab Technical University, Jalandhar, in 2017. He has more than 17 years of teaching experience. His research interests are IoT, machine learning, and natural language processing. He has authored many published books with reputed publishers. He works as a reviewer for several journals, including *Big Data*, *CMC*, *Scientific Report*, and *TSP*. He has authored or co-authored more than 50 publications at various national and international conferences and in international journals.

Chadi Altrjman is Associate Scientist and Researcher at Near East University, Cyprus, and has earned a bachelor of engineering at the University of Waterloo, Ontario, Canada. He is involved in several smart projects by which he is expanding his practical skills. His research interests include AI and machine learning, wireless sensor networks, chemical sensors, sustainable and renewable energy integration, distributed power generation, and smart grid applications.

Contributors

Hassana Abubakar
Near East University
Cyprus

O. F. Adebayo
University of Ibadan
Nigeria

Shweta Agarwal
Chandigarh University
India

Shakeel Ahmed
King Faisal University
Saudi Arabia

Aishwarya M
Jyothy Institute of Technology
India

S. M. Akintunde
Federal University of Agriculture
Nigeria

O. V. Alomaja
Yaba College of Technology
Nigeria

Fadi Al-Turjman
Near East University
Cyprus

Sinem Alturjman
Near East University
Cyprus

Zubaida Sa'id Ameen
Near East University
Cyprus

Satyabrat Malla Bujar Baruah
Tezpur University
India

Bidyut Bikash Borah
Tezpur University
India

Sharad Chauhan
Chitkara University Institute of Engineering and Technology
Chitkara University, Punjab
India

Dimple Chehal
Manav Rachna University
India

Atharva Deshmukh
Terna Engineering College
India

Gokul H
Vellore Institute of Technology
India

Payal Gulati
J. C. Bose University of Science and Technology, YMCA
India

Meenu Gupta
Chandigarh University
India

Harshitha M V
Jyothy Institute of Technology
India

Shraddha Jathar
Terna Engineering College
India

Debaraj Kakati
Tezpur University
India

Deepti Kakkar
Dr. BR Ambedkar National Institute of Technology
India

Kusrini
University of Amikom
Indonesia

Auwalu Saleh Mubarak
Near East University
Cyprus

Nisha
GCW Behal
India

A. O. Okewale
University of Ibadan
Nigeria

T. O. Olaleye
Nigeria Computer Centre and Services
Federal College of Education
Nigeria

Kanika Pahwa
SRM University
Delhi, India

Nitika Phull
I. K. Gujral Punjab Technical University
India

Chander Prabha
Chitkara University Institute of Engineering and Technology
Chitkara University, Punjab
India

Soumik Roy
Tezpur University
India

Nida Shafin
Jyothy Institute of Technology
India

Neha Sharma
Chandigarh Group of Colleges
India

Jaspreet Singh
Chandigarh University
India

Joy Karan Singh
Dr. BR Ambedkar National Institute of Technology
India

Parminder Singh
Chandigarh Engineering College
India

Swati Singh
CSIR-National Botanical Research Institute
Academy of Scientific and Innovtive Research (AcSIR)
India

I. Solanke
Yaba College of Technology
Nigeria

Srushti J
Jyothy Institute of Technology
India

Harshavardhan Tiwari
Centre for Incubation, Innovation, Research and Consultancy (CIIRC)
India

Tushita S
Jyothy Institute of Technology
India

Amit Kumar Tyagi
Vellore Institute of Technology
India

Amit Verma
Chandigarh University
India

Chapter 1

Introduction to computational intelligence in healthcare

Applications, challenges, and management

Chander Prabha, Jaspreet Singh, Shweta Agarwal, Amit Verma, and Neha Sharma

CONTENTS

1.1 Introduction 1
1.2 Fundamentals of CI 3
1.3 Computational intelligence techniques in healthcare 6
1.4 Case studies in the use of CI techniques in improving patient outcomes 9
1.4.1 Case study: Cancer diagnosis with AIML 9
1.4.2 Case study: Hospital/patient safety improvement with operational/integrated data analysis 10
1.5 Ethical implications of CI in management and challenges 12
1.6 The perspective of CI in the future and applications 13
1.7 Conclusion 13
References 14

1.1 INTRODUCTION

Artificial intelligence plays an important role in the healthcare industry. However, human expertise and new techniques comprising computational intelligence methods, viz. fuzzy models, evolutionary algorithms, artificial neural networks, probabilistic methods, genetic algorithms, and swarm intelligence, emerged as the most promising tools in developing intelligent systems in healthcare and its applications.

Computational intelligence (CI) is a subset of AI. Real-world problems which requires decision-making and reasoning can be solved using CI [1]. The intelligent system can be designed using CI to have capabilities of human-like intelligence in human-made machines. CI is efficient and produces tractable solutions, and is simpler and more robust than traditional approaches.

The recent technological advances in CI have shown very encouraging results while doing statistics computation (data analytics) in the field of

DOI: 10.1201/9781003305347-1

healthcare, social media studies, industry, and business, and pattern recognition [2]. Figure 1.1 shows a relationship between AI and CI [2]. AI comprises an expert system, NLP, machine vision, speech recognition, and many more, whereas CI comprises neural networks, genetic algorithms, evolutionary algorithms, fuzzy logic, swarm intelligence, and more.

The recent developments in the field of AI, machine learning (ML), and big data have increased the importance of signal processing and biomedical image research. Biomedical image processing and biomedical signal processing are similar in concept in many ways. It includes the analysis, presentation, and enhancement of images captured via X-ray, nuclear medicine, MRI, and visual imaging technologies [3]. ML is now rapidly being used in all engineering science research fields, especially in healthcare, including biomedical sciences. In a similar context, CI paradigms comprising numerous branches, including swarm intelligence, neural networks, evolutionary computing, expert systems, artificial immune systems, and fuzzy systems, play a vital role in dealing with different aspects of pattern recognition and data analytics in healthcare. With the help of the previously mentioned techniques, CI tries to predict the human nature of questioning and reasoning to build an intelligent machine. Table 1.1 shows the difference between CI and AI.

Nowadays, advanced techniques help in understanding the human body and processes, and handling health data anywhere and anytime. The maintained healthcare system meets the problems and challenges that arise in the healthcare industry comprising patients, doctors, disease monitoring/

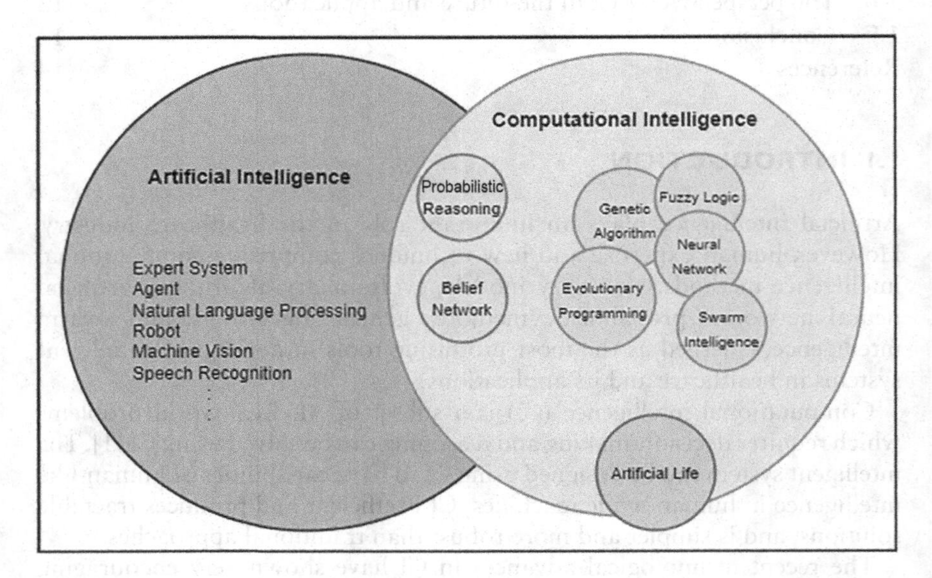

Figure 1.1 Relationship between CI and AI [2].

Table 1.1 Difference between CI and AI

Computational intelligence	*Artificial intelligence*
Uses soft computing technique	Uses hard computing technique
Follows fuzzy logic	Follows binary logic
Uses nature-inspired models	Is based on mathematical models
Can work on inexact and incomplete data	Is not very effective
Provides probabilistic results	Provides deterministic results

prevention, hospital management, decision-making modules, and many more. Furthermore, the advanced healthcare systems must be upgraded with new capabilities to provide mankind with more professional and intelligent healthcare services. The next section describes the fundamentals of computer intelligence, along with its associated techniques and procedures in healthcare.

1.2 FUNDAMENTALS OF CI

Initially, AI has been extensively used in many industrial applications. Specifically, AI deals with the study of computational agents in their synthesis and analysis. In AI, a computational agent acting in an environment means it is doing something. However, in CI, a computational agent is an agent whose decisions about tasks/actions can be elaborated in terms of computation. When talking about the research area of CI in healthcare, it mainly comprises paradigms, concepts, algorithms, and implementations to develop new machines/systems that exhibit intelligent behavior to predict correct results in complex environments. The role of CI in healthcare has its own significant contributions [4]. The adopted general strategy in the field of CI is to apply approximation methods and techniques that can find coarse or only partially valid solutions to given problems. The area of CI is vast. This section describes various possible CI techniques used in healthcare.

Figure 1.2 shows the classification of computational techniques [11]. Intelligent machines can be developed by both AI and CI, but they are different. The top-down approach is followed in AI, while CI follows a bottom-up approach. Nowadays, CI is used heavily in biomedical healthcare, apart from its role in engineering, data analysis, and so on. The main paradigms of CI according to Engelbrecht [5] are as follows:

- Artificial neural networks (ANN)
- Evolutionary computations (genetic algorithms)
- Swarm intelligence (ant colony, particle swarm, etc.)
- Fuzzy systems
- Learning theory
- Probabilistic methods

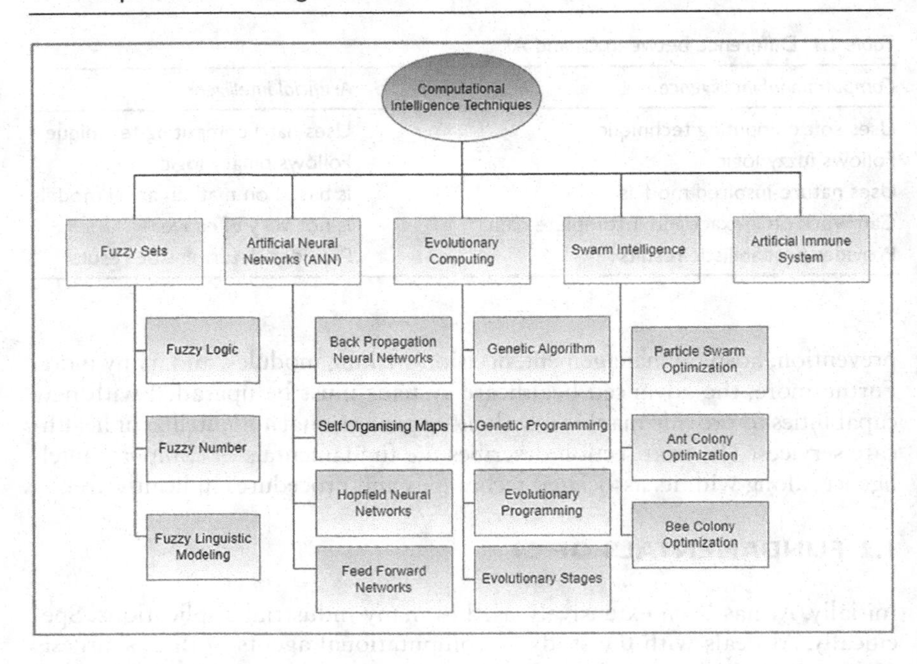

Figure 1.2 CI techniques classification [11].

Major advancements in the field of neural networks in terms of artificial neural networks (ANN) are based on biological studies. The elements of ANN are the cell body, axon, and synapse. The cell body's task is to process the information, the axon is a device that enables signal conduction, and the synapse controls those signals. The ANN basically learns from experimental data. Furthermore, ANN can be combined with fuzzy to get the advantage of data clustering. The principles and ideas in the field of ANN are motivated by neuroscience. Basically, ANNs are information processing systems (IPS). The function and structure of ANN simulate the human/animal nervous systems, particularly the brain. The ANN consists of a large volume of fairly simple processing units working in parallel, called neurons. These neurons further send information in the form of action via directed links to other neurons. To achieve the learning capability, the CI tries to model and mimic the behavior of biological neural networks [6].

The idea behind evolutionary algorithms [7] eventually comes from the study of biological evolution. Here, over many generations, organisms adapt per environmental conditions. Certain classes of optimization problems are addressed by evolutionary algorithms. These algorithms belong to the metaheuristic family, which further offers algorithms to approximately solve many types of optimization problems. The metaheuristics consist of an abstract sequence of steps, and every single step implementation must be in a defined problem-specific fashion. These are often applied to solve problems for which no optimal/efficient solution exists. In the earlier cases, biologically inspired methods, such as PSO (particle swarm optimization or computational swarm intelligence), have proven themselves to be majorly implemented in healthcare theory and practice.

In fuzzy systems, the vague information provided either by a system developer or a human expert is formalized using fuzzy set theory and fuzzy logic (FL). The use of the FL technique provides intelligent, knowledgeable information along with reasoning to handle uncertainty, inaccuracy, and vagueness. Due to the previously mentioned capabilities, these systems have been applied successfully in healthcare to describe the behavior of complex healthcare systems. Fuzzy logic–based algorithms imitate human thought processes in the crucial sphere and precisely execute repetitive tasks. These tasks, however, are ill-suited for humans if done repeatedly. In general, for any medical condition, there may or may not be chances for the existence of a precise model, but if it exists, then it is too troublesome to model. In such scenarios, FL is considered to be a valuable tool, as the mind of humans works on approximate data. Fuzzy logic helps in extracting meaningful information and producing crisp solutions [8]. The human's expert knowledge in using linguistic terms is transferred into manageable models using fuzzy logic. A number of if-then-else rules are used in the FIS (fuzzy inference system) of fuzzy logic. These rules are based on the knowledge extracted from experts in the field to assist in diagnosing any disease and prescribing the necessary medical treatment for the detected symptoms [9].

Apart from the previously described, more advanced techniques based on learning theory and probabilistic methods [10] are the new ones used in today's healthcare CI techniques. In learning theory, one looks for a way that is close to the humans' (i.e., reasoning). Learning theory helps to make predictions based on previous experience by understanding surrounding environments' effects and experiences. The probabilistic method is considered as one of the main elements of FL, whose aim is to evaluate the results of a CI system, defined by a randomness function. Hence, probabilistic methods based on prior knowledge bring out possible solutions to a problem.

Increased healthcare needs result in an increase in health monitoring systems for the aged population. The remote monitoring approach proved to be most cost-effective unless and until an extreme emergency arises. Wireless patient monitoring systems equipped with one or more CI techniques are available based on long-range and short-range communications protocols, such as GPRS, Zigbee, RFID, and Bluetooth. Table 1.2 shows the strength and weaknesses of CI techniques [12].

Table 1.2 Strengths and weaknesses of CI techniques [12]

CI technique	*Strength*	*Weaknesses*
Fuzzy logic	• Uncertainties and ambiguities can be identified effectively. • Fuzzy product similarities are what it deals with.	• Continuous interaction with experts in the field is required to resolve ambiguities. • Diversified user needs make it difficult to combine outputs from multiple rules.
ANNs	• Similar behaviors of users can be effectively identified. • Context elicitation is accurate and effective.	• Explanation of results and interpretation is limited.
EC	• Context computation is accurate. • Query optimization, document search, and personalization are effective uses for it.	• Convergence is premature. • Computation is expensive.
SI	• Highly evolvable and adaptive. • Less memory is required.	• Problems that require an exact solution is not meant for this. • Absence of centralized control means SI falls into inactivity.
AIS	• Continuous learning is supported and immunological memory is maintained of new and past encounters, respectively. • Noise from web contents can be removed.	• Adaptation of profile in practice is challenging. • Naive approaches are required to extract traditional metaphors.

1.3 COMPUTATIONAL INTELLIGENCE TECHNIQUES IN HEALTHCARE

CI techniques are accustomed to medical data. Healthcare is an all-around realm incorporating remote monitoring, advanced decision-making, operational excellence, healthcare logistics, and present-day information systems. Work is in progress in identifying and investigating the use of computational techniques to address the complexity and scale of the problems in healthcare to make it smart and informative. This assimilation will be handy in making a hospital into a very smart and next-generation revolutionary hospital with a complete digital healthcare ecosystem by using CI methodology, as CI can adapt itself according to the problems specific in the domain.

In healthcare, medical data [13] is divided into certain categories based on its attributes. It may be one form or in combination, like narrative or textual medical data, viz. family or social history, or descriptions or answers provided by the patient; numerical measurements, viz. lab results and other observations; recorded signals, viz. ECG, radiologic images, and graphical tracing pics of CT scan images; and heterogeneous data. Figure 1.3 shows medical data categories. Medical data serves as input for the various CI techniques during implementation [14].

The source of data and its maintenance resource person preserve the medical data in the form of records, viz. EMRs or medical records stored in electronic form or as digital formats of the information and notes, to provide care to patients. One repository in the form of EHRs (electronic health records) is maintained that contains information from all the doctors/ clinicians involved in a patient's care, and they all can access the information available in the EHR of a patient. Another type of record is PHRs (personal health records) that contain the same information as EHRs but are designed to be accessed by patients themselves. Figure 1.4 shows the historical development of the AI boom in clinical applications over time in the form of diagnosis, treatment, and follow-up.

Table 1.3 shows the trigger, trends, and problems in the AI boom [15]. In the future, CI, in combination with AI [16], will play an important role in healthcare. Early efforts to provide treatment and diagnosis using CI have proven challenging, but soon, CI will capture the whole healthcare market. Say, with the fast evolution in AI for imaging analysis, most pathology and radiology images will be examined at some point by a machine.

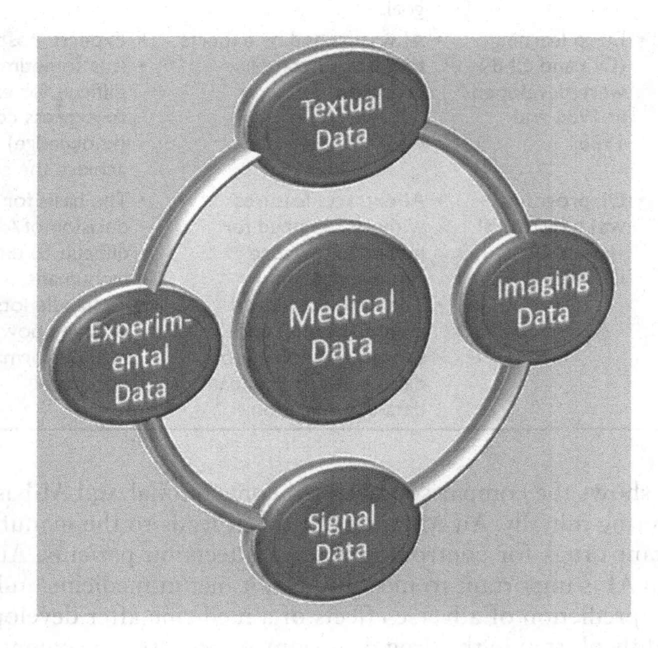

Figure 1.3 Medical data categories.

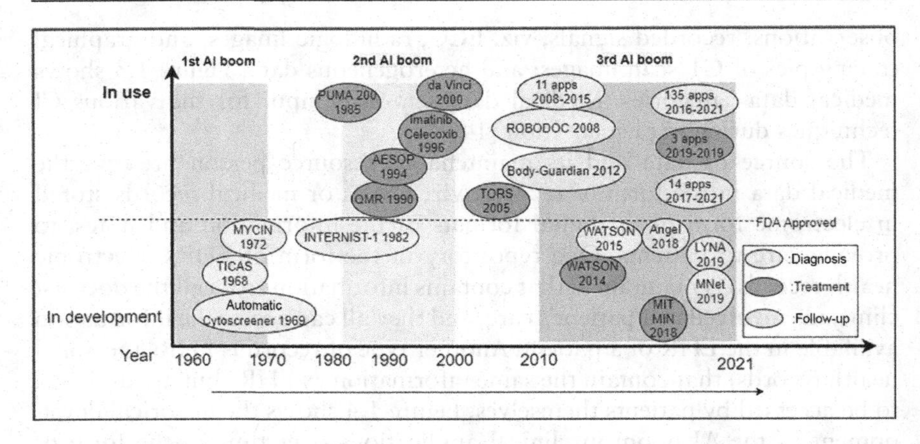

Figure 1.4 A historical view of clinical applications with computational support [15].

Table 1.3 AI boom trigger, trends, and problems [15]

AI	*Trigger*	*Trends*	*Problems*
1st boom	• AIML was proposed in 1959.	• It searched for the purpose (goals) by segregating cases per different paths towards goal.	• It is difficult to solve complex problems.
2nd boom	• Deep learning (DL) and CNN were developed in 1986 and 1988.	• AI was guided by experts to adapt to complex problems.	• Experts' cost was high. • It is sometimes difficult for experts to express conditions (knowledge) to achieve the goal.
3rd boom	• DL program was a historical victory in ILSVRC, 2012.	• AI extracts features without the need for human knowledge representation • The development of communication and sensor technology (big data) leads to abundant data accumulation.	• The basis for the decision of AI is difficult to understand by humans. • To handle lots of data, machine powered with high performance is necessary.

Table 1.4 shows the comparison between conventional and AI-based drug manufacturing initially. An AI-based approach leads to the manufacturing of important drugs for controlling disease effects on patients. Automatic learning in AI is important to meet needs not met in medicine and life sciences. The prediction of adverse effects of a medicine after development is the most difficult step in the drug development process. The computational modelling technique using ML is useful for predicting adverse effects [17].

Table 1.4 Comparison between conventional and AI-based drug manufacturing

Development of drug	*Conventional*	*AI-based*
Driving factor	• Target-driven	• Data-driven
Target	• Known cell structure and interaction among cells lead to easily druggable targets.	• Targets are extracted from ML using big data.
Advantage	• Humans can easily understand the identified targets.	• It is done before synthesis by predicting the properties of compounds and activities.
Problems	• The complex, unknown interactions and structure of compounds lead to limited targets. • Promising compound identification is time-consuming.	• It is necessary to learn from a large amount of accurate data. • The AI cannot understand whether the compounds are meaningful to humans or not.

1.4 CASE STUDIES IN THE USE OF CI TECHNIQUES IN IMPROVING PATIENT OUTCOMES

Innovative ways are necessary for providing more proactive care and for reducing the pressure on hospital management/services with the growing and aging population. This section describes various case studies showing the use of CI techniques.

1.4.1 Case study: Cancer diagnosis with AIML

In this case study, the research was funded by the UK government and is given the name AID-GI Project. A study was done to improve gastrointestinal (GI) diseases and colon cancer using AIML. The traditional state-of-the-art endoscopy procedure uses tube-mounted cameras, whereas endoscopy is now done using the revolutionary video camera capsule (built using CI techniques shown in Figure 1.5) that offers a much more significant advantage. The major highlights of the invention are the about 400,000 images it can produce during its journey into a patient's body and the reduction of up to 56% of the time spent by a specialist for diagnosing. Nearly 45,000 colonoscopy procedures are carried out in England every year.

The challenge here is the volume of images taken by the capsule, which requires consistent high-quality analysis and is not achievable with manual methods. Here, machine learning–based automated image analysis is used, thus, removing the need for data processing. The analysis is done with the help of the CapScan tool. Images can be imported, and videos can be visualized at a simple interface (Figure 1.6). Thousands of images can be imported

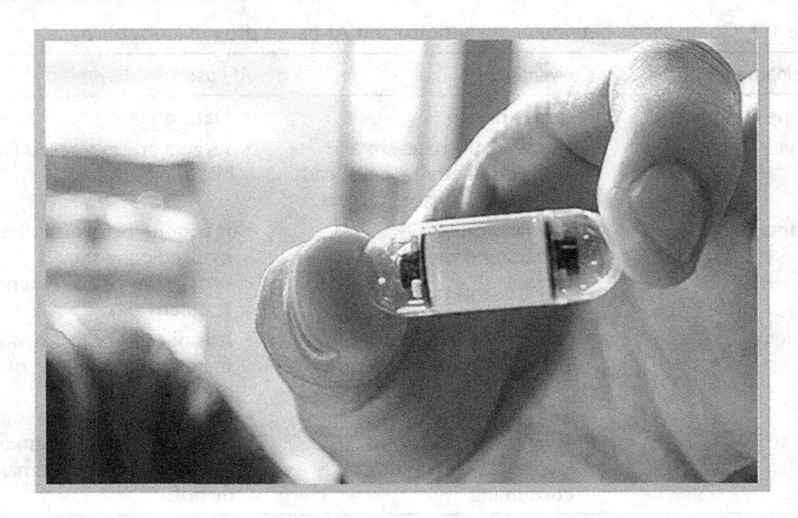

Figure 1.5 Video capsule (pill-size) [18].

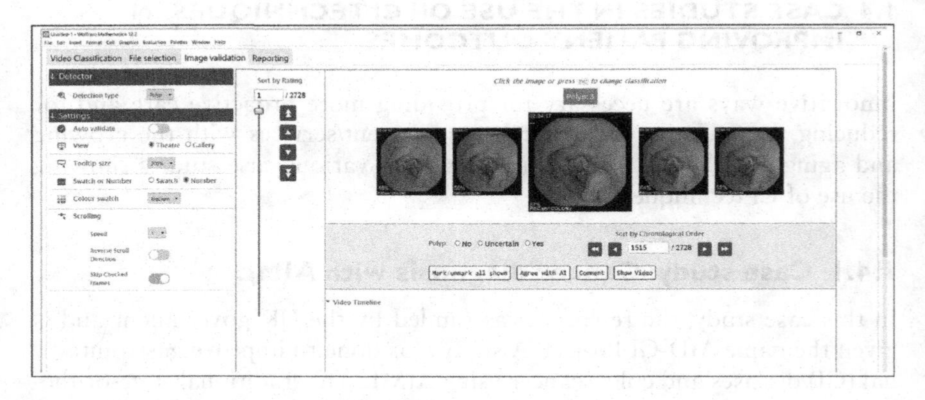

Figure 1.6 CapScan tool image [19].

by clinicians, and they can automatically identify any abnormalities. The achievements of the GI tract analysis tool are listed in Table 1.5.

1.4.2 Case study: Hospital/patient safety improvement with operational/integrated data analysis

This case study [20] is with respect to the National Health Service (NHS) England, having over 1.5 million employees and catering to 1 million patients every 36 hours. Every single day, huge data is being generated across a range

Table 1.5 GI tract analysis tool achievements

	Achievements	*Description*
1	Enabling health tech innovations	Now, the operators of the CapScan tool do not need to search thousands of individual images of a single patient. They are now able to quickly select an image and can diagnose the symptoms with CI matching image.
2	Reducing the burden on hospitals and improving healthcare	The GI tract analysis is quick and accurate, with no need for specialized consultants and doctors, and can be done in community healthcare centres.
3	Providing scope for continuous improvement	The verified images are added back as a training set to make the tool more reliable, thus improving its performance as well.

of hospitals, clinical care centres, and health trusts that is disseminated and to be collected from and over multiple disparate databases. This makes it difficult to utilize data effectively. The Wolfram consulting team helps in importing, cleaning, and analysing the data quickly, and pairing collected data with nearly 1,000 parameters. The achievements are as follows:

- Using integrated data analysis, the Wolfram team uncovered hidden signals to provide clear and actionable insights to improve hospital safety.
- Making use of the rapid analysis technique to deliver results in just ten days.
- Doing investigations with cutting-edge technology, along with a multiparadigm approach.

Figure 1.7 shows an approach for improving patient safety given by Professor Patrik Hudson [21]. The improvement in patient safety focuses on three parameters: describing a safety issue (identifying and measuring), intervening (acting to help the patient), and preventing (avoiding the same type of events in the future). The approach given by Professor Patrik combines technology, process, and culture, and was named the sociotechnical approach. To increase awareness and trust, the ladder score is defined from very low to very high, indicating the culture of safety in regard to health system performance. In Figure 1.7, the culture encapsulates patient care, like family, teamwork, leadership, and the economic impact of culture. The process encompasses reliability, organizational fairness, and process improvement. The technology includes IT healthcare using CI techniques.

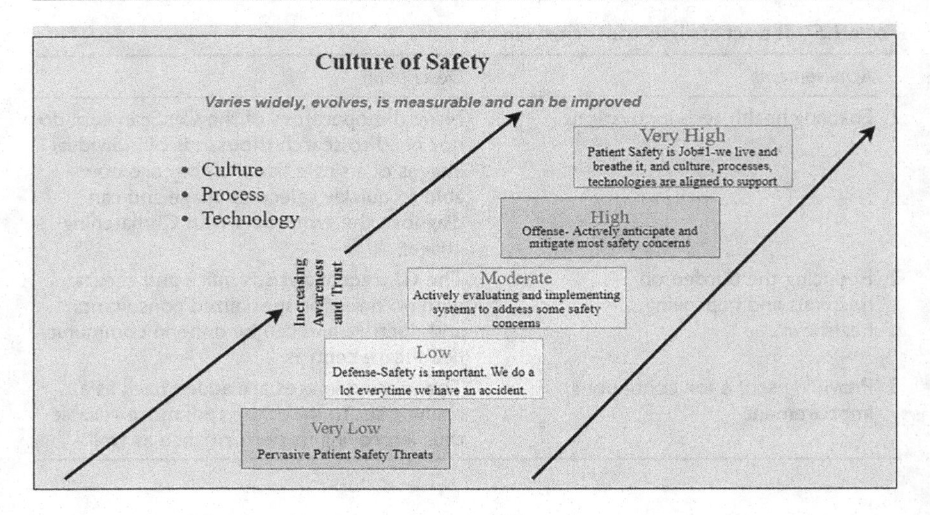

Figure 1.7 Sociotechnical approach for improving patient safety [21].

1.5 ETHICAL IMPLICATIONS OF CI IN MANAGEMENT AND CHALLENGES

In today's technology, transparency is perhaps the most difficult issue to address. Many AI and CI techniques, mainly deep learning (DL) algorithms, are used for image analysis. Sometimes, it seems impossible to interpret or explain an image captured. For example, if image analysis is to be done of a patient diagnosed with cancer, there may be a case when either the DL algorithms or the physician is unable to provide an answer to the patient's question about why it happened to them. There may be chances of mistakes happening with the tools and techniques employing CI, and hence, it may be difficult to establish accountability if a patient is diagnosed with cancer per his image analysis, then either the physicians or DL algorithms are unable to provide an explanation in regard to why [22].

Changes need to be done in medical, technological, and ethical standards in the future, and proper structures need to be established, and governmental and regulatory bodies need to monitor the key issues in healthcare institutions [23]. To reduce the negative implications, the regulatory bodies must act in an answerable manner. The diversity of CI's applications is their us in, viz., autonomous vehicles, factory processes scheduling, chess-playing, natural language translation systems, medical diagnosis, and much more. Thoughtful policy mechanism and continuous attention are required to implement these technologies in practice for use.

1.6 THE PERSPECTIVE OF CI IN THE FUTURE AND APPLICATIONS

Analysis is an important aspect, per researchers, of the overall amount a particular healthcare organization pays for the use of CI technologies. If the use of CI technologies is necessary for patient treatment and health service management, then the regulatory bodies controlling and managing healthcare should contribute and invest in the modernization of healthcare. The investment funds could be available as a part of the national investment program in the healthcare world, just as how it was done by the European Union in the Next Generation EU program [24]. The poorest countries in the world lack services and infrastructure with respect to health and medicine and would be in need of more medical funds [25]. Additional profits generated could be evaluated for comparison among those healthcare organizations that don't use CI technologies versus those that use them. Another analysis can be done to identify the reason why some parts of the world have not done studies in this field. Comparative analysis can be done. Research can also be done to identify the factors affecting CI technology implementation. The collaboration among countries presents ahead a challenge with respect to the exchange of ideas and innovative use of CI technologies in healthcare.

In other fields, viz. business, accounting, and management research, currently, there is a lack of quantitative analysis of the profits and costs emerging from healthcare organizations that use CI technologies. Finally, more interdisciplinary studies are needed to strengthen the use of CI to improve healthcare with quality data management, along with ethical considerations in healthcare [26].

1.7 CONCLUSION

A lot of research has been done for decades on the alignment of efficient and futuristic techniques in computational intelligence from the beginning, keeping in mind the academic environment, along with social domains in healthcare. These technologies, over the last years, experienced an industrial evolution in making intelligent devices, providing society with a reliable change in accuracy. Mainly, in the healthcare practice and the medical industry, these technologies and tools play an important role. The CI techniques used in healthcare have been discussed in this chapter. Case studies have been presented, along with ethical implications and future use perspectives of CI. One can further think about the collaboration of CI techniques for improving healthcare facilities in communities with greater accuracy and, thus, reducing the burdens on hospitals in the healthcare industry.

REFERENCES

[1] Wiwanitkit, S., Wiwanitkitn, V., "Computational intelligence in tropical medicine", *Asian Pacific J. Trop. Biomed.*, 6(iv), pp. 350–252, 2016.

[2] Hernandez-Julio, Y.F., et al., "Framework for the development of business intelligence using computational intelligence and service-oriented architecture", in *Proceedings of the 12th Iberian Conference on Information Systems and Technologies* (CISTI), pp. 1–7, 2017.

[3] "Computational intelligence", *Wikipedia, the free encyclopedia.* Available from: https://en.wikipedia.org/wiki/Computational_intelligence.

[4] Kacprzyk, J., Pedrycz, W. (eds.), *Springer Handbook of Computational Intelligence*, Berlin/Heidelberg: Springer, 2015.

[5] Engelbrecht, A.P., *Computational Intelligence: An Introduction*, 2nd Edition, Chichester, West Sussex: John Wiley & Sons, 2007.

[6] Bredell, G., Tanner, C., Konukoglu, E., "Iterative interaction training for segmentation editing networks", *Comput. Sci., ArXiv, abs/1807.08555*, 23rd July 2018.

[7] Vikhar, P.A., "Evolutionary algorithms: A critical review and its future prospects", in *International Conference on Global Trends in Signal Processing, Information Computing and Communication (ICGTSPICC)*, pp. 261–265, 2016.

[8] Mahfouf, M., Abbod, M., Linkens, D., "A survey of fuzzy logic monitoring and control utilization in medicine", *Artif. Intell. Med.*, 21, pp. 27–42, 2001.

[9] Garibaldia, J., Zhou, S.M., Wang, X.Y., et al., "Incorporation of expert variability into breast cancer treatment recommendation in designing clinical protocol guided fuzzy rule system models", *J. Biomed. Inform.*, 45, pp. 447–459, 2012.

[10] Hommersom, A., Verwer, S., Lucas, P.J.F. "Discovering probabilistic structures of healthcare processes", in Riaño, D., Lenz, R., Miksch, S., Peleg, M., Reichert, M., ten Teije, A. (eds.) *Process Support and Knowledge Representation in Health Care. ProHealth KR4HC 2013.* Lecture Notes in Computer Science, vol. 8268, Cham: Springer, 2013. https://doi.org/10.1007/978-3-319-03916-9_5.

[11] Sadiku, M.N.O., Foreman, J., Musa, S.M., "Computational intelligence", *Eur. Sci. J.*, 14(21), ISSN: 1857-7881 (Print) e-ISSN 1857-7431, July 2018.

[12] *A Survey on Context-aware Recommender Systems Based on Computational Intelligence Techniques.* Available from: https://www.researchgate.net/publication/272508863_A_Survey_on_Context-aware_Recommender_Systems_Based_on_Computational_Intelligence_Techniques

[13] Dash, S., Shakyawar, S.K., Sharma, M., et al., "Big data in healthcare: Management, analysis, and future prospects", *J. Big Data*, 6, p. 54, 2019.

[14] Kalantari, A., et al., "Computational intelligence approaches for classification of medical data: State-of-the-art, future challenges and research directions", *Neurocomputing*, 276, pp. 2–22, 2017.

[15] Asai, A., Konno, M., et al., "Computational healthcare: Present and future perspectives (review)", *Exp. Therap. Med.*, 22(6), 2021.

[16] American College of Radiology. *Cleared AI Algorithms*, 2021. Available from: https://models.acrdsi.org. Accessed 1 July 2021.

[17] Jamal, S., Ali, W., Nagpal, P., Grover, S., Grover, A., "Computational models for the prediction of adverse cardiovascular drug reactions", *J. Transl. Med.* 17(171), 2019.

[18] Innovate UK, "Result of competition: Precision medical technologies: shaping the future", November 2018. Available from: https://www.wolfram.com/customer-stories/revolutionising-cancer-diagnosis-with-ai-and-machine-learning.pdf
[19] https://www.wolfram.com/computational-intelligence-in-healthcare/
[20] https://www.england.nhs.uk/wp-content/uploads/2020/08/190708_Patient_Safety_Strategy_for_website_v4.pdf
[21] https://www.healthcatalyst.com/insights/use-data-improve-patient-safety
[22] Gupta, M., Gupta, A., Sandhu, G., "Deep learning for medical healthcare: Issues, challenges, and opportunities," in *Handbook of Computational Analysis and Understanding of Deep Learning for Medical Care: Principles, Methods, and Applications*, Wiley-Scrivener (CUDLMC: 2020), ISBN-13: 978-1119785729, 28th July 2021.
[23] Tyagi, A.K., Gupta, M., Ved, C., "Healthcare solutions for smart era: A useful explanation from user's perspective," in *Recent Trends in Blockchain for Information Systems Security and Privacy*, 1st Edition, CRC Press, ISBN 9780367689438, 22 November 2021, eBook ISBN9781003139737.
[24] Picek, O., "Spill over effects from next-generation EU", *Intereconomics*, 55(5), pp. 325–331, 2020.
[25] Sousa, M.J., Dal Mas, F., Pesqueira, A., Lemos, C., Verde, J.M., Cobianchi, L., "The potential of AI in health higher education to increase the students' learning outcomes", *TEM J.*, 10(2), pp. 488–497, 2021.
[26] Secinaro, S., Calandra, D., Secinaro, A., et al. "The role of artificial intelligence in healthcare: A structured literature review", *BMC Med. Inform. Decis. Mak.* 21, p. 125, 2021.

Chapter 2

Role of IoT and machine learning for e-healthcare management

Nitika Phull, Parminder Singh, and Kusrini

CONTENTS

2.1 Introduction 18
 2.1.1 Applications of IoT in healthcare 19
2.2 Background study 21
2.3 IoMT (internet of medical things) 23
 2.3.1 IoMT applications 24
 2.3.2 IoMT components 25
2.4 Non-wearable sensors 27
 2.4.1 Smart pill box 27
 2.4.2 Smart beds 27
 2.4.3 Human activity detection 27
2.5 Personalized medicine using IoT and ML 28
 2.5.1 Personalized healthcare 28
2.6 Application and challenges of IoT in healthcare 30
 2.6.1 Challenges of IoT in healthcare 30
2.7 Benefits of IoT in healthcare 30
 2.7.1 IoT for doctors 31
 2.7.2 IoT for patients 32
 2.7.3 IoT for hospitals 33
 2.7.4 Application scenario 34
2.8 Application of machine learning in the healthcare industry 35
 2.8.1 Drugs discovery 36
 2.8.2 Cancer diagnosis and prognosis 37
 2.8.3 Medical imaging 38
 2.8.4 Smart health records 39
  2.8.4.1 Arduino fio 40
2.9 Disadvantages and challenges of IoT and ML in healthcare 41
2.10 Conclusion and future scope 42
References 43

DOI: 10.1201/9781003305347-2

2.1 INTRODUCTION

The healthcare and technology sector have becomes extremely involved and interconnected over time. Moreover, IoT is gradually booming in almost all zones of healthcare. The success of mini devices which can be worn and the public adoption of AI and machine learning (ML), along with big data analytics, are rising quickly, delivering exciting personalized treatment, prospects and special insight facilities. This tremendous new wave of IoT-based creativity is capable of extending the healthcare limits outside the hospital settings during the transition from hospital-centred to patient-centred care from the ecosystem. Psychological signs, vital signs and health-related information of the subjects are consistently monitored in such a system at any location by various biosensors from anywhere. Further, this information is made available to healthcare providers and other qualified groups, usually through the IoT cloud platform.

The implementation of cloud computing and the cloud established a model, the cloud-based IoT, partially to solve some of IoT's main difficulties, like minimal functionality and storage. Cloud computing is a computing paradigm for universal, efficient and on-call facilitation, including database, storage, application and services, to a common pool of configurable computing resources. Many companies have also begun to substitute conventional business intelligence (BI) methods for analysing health data over the past few years. In this context, businesses have begun to leverage artificial intelligence, especially machine learning, in their application for e-health.

IoT is the main area of research in the healthcare community in the past years. The healthcare industry is important, and IoT has many options to make it much better. Many medicinal gadgets and sensors can be attached over the network, which enables access to significant information about patient scenarios. The information may then be used for manifold scenarios, such as monitoring patients from remote locations, guessing recovery and illness rates through better inspection into symptoms, and generally refining the diagnosis and treatment process through more automation. Because of the high volume of data produced in daily life scenarios by such devices and their compounding nature, analysing and interpreting data using ML procedures are of great importance to healthcare IoT. Such algorithms make important inferences from the gathered data and make valuable interpretations. Moreover, machine learning models can provide good accuracy when trained in the perfect way. Many advanced research are targeted at searching fresh domains where ML algorithms can be applied in the healthcare IoT systems, evaluate the suitability of the systems, and enhance the correctness achieved by ML models [31].

2.1.1 Applications of IoT in healthcare

Due to the advancement of technologies, IoT has proven to be the best for the medical industry. IoT allows the creation of a network of devices which can generate and exchange data in the network framework. IoT has transformed regular hospitals into the smart hospitals. With the advancement of technology, data is collected in a centralized database, and everything is tracked simultaneously. IoT enables us to monitor the health of patients remotely while the patients are sitting at home. Some of the applications of IoT are as follows:

- **IoT for patients:** IoT has changed people's lifestyles by keeping track of their health issues. For example, with the invention of wearables, like exercise bands and other related devices, the daily activities of a person can be monitored, and an automatic alarm system will send a warning message to family members and keep them informed about his health conditions. IoT devices offer patients personalized care. These machines can be designed to alert you of things like exercise schedules, appointments with the doctor, blood pressure changes and calories.
- **IoT for hospitals:** IoT can be used in various medical instruments, like life ventilators, oxygen supplies, heart monitors and nebulizers. Many intelligent devices, that is, IoT-based devices, monitor the health of the patient all day long. IoT helps the healthcare industry take care of patients beyond the boundaries of hospitals.
- **IoT for medical practitioners:** IoT helps medical practitioners check on a patient's health by having the patient use wearable devices and other IoT-enabled home investigation devices. It helps them check whether patients are sticking to their recovery timetables or if they require some special medical care. IoT-enabled devices allow healthcare stakeholders to be more watchful and concerned of their patients.
- **IoT for insurance companies:** With the advancement in the IoT industry, this area helps medical insurance companies obtain data regarding the health status of any patient. Based on the health tracking device, they can handle all the claims. This will enable companies to manage healthcare claims in given timelines. Also, it will help them detect any fraudulent and fake medical claims.

There are numerous benefits of IoT in the health industry. It helps analyse the patient's data with fewer errors. It can also help medical practitioners manage medical and drug equipment. The data of the patient can be collected efficiently and with fewer error with the help of IoT devices. Having IoT will bring more alertness among the medical staff members as they monitor the health statistics of the patient in real time simultaneously. With IoT, the medical industry can perform with efficiency and with less cost in the long run.

Machine learning techniques are strongly dependent on the machine's computing power, particularly in the training phase. In this respect, the cloud enables advanced algorithms to easily write observations and dynamically recognize irregularities from the vast quantities of unstructured and structured data generated over time by IoT devices. In the IoT, e-health, cloud-oriented model, the tremendous amount of information has to be transmitted to the cloud from computers, and it needs a large amount of accessible message with the bandwidth. Moreover, cloud-oriented storage requires considerable redundancy. This is because of the information sharing in the cloud between servers and computers. The cloud-based IoT model is unable to fulfil the high processing time requirements in latency-sensitive applications requiring service in real time. A prime example of this in the situation of e-health are the arrhythmia control and the categorization of volume, variation, and pace, and also the respond time and message bandwidth that can easily be dealt with. The leveraged approach, therefore, tries to respond to any adjustment on the fly irrespective of the accessibility of tracking devices, considering the extremely complex and real-time existence of e-health software. Fog or edge computing has recently arose as a method to fix the disadvantages of e-health models of cloud-based IoT. Through positioning service nodes close to the data source at the network edges, edge computing improves the conventional e-health architectures.

This design helps us maintain the real-time reactivity required by e-health programs by decreasing the latency that comes up during the reading move between the IoT and cloud endpoints. A new development in IoT is the implementation of ultra-low AI-based accelerators with controls that can conduct machine learning on even a device or a chip. In the healthcare sector, particularly, in this chapter, we suggest a modern multi-layered structure based on the collaborative machine paradigm for IoT e-health applications. We analyse and evaluate traditional architectures of IoT e-health and explore how to accurately exploit device-edge-cloud capabilities to achieve a low consumption of electricity and low latency. Linear discriminant (LD) algorithms, AdaBoost, genetic algorithm back propagation neural network (GA-BPNN), multi-layer perception (MLP) propagation and support vector machine (SVM) were successfully used to analyse the ECG signals. None of these are fully tailored to be used in IoT e-health applications to the best of our understanding and can, therefore, not address the significant limitations of IoT solutions, such as limited resource processing and restricted battery life [1]. Figure 2.1 shows how IoT can be used in the healthcare industry.

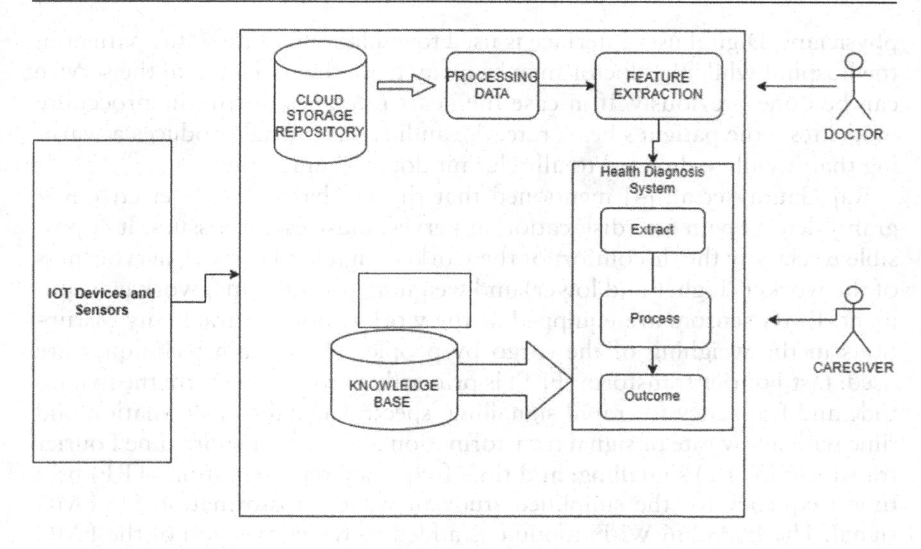

Figure 2.1 IoT in the healthcare industry.

2.2 BACKGROUND STUDY

Claudia C. et al. [2] have monitored e-health applications structured with the quality of research results, analysis techniques and procedure algorithms, which are a few methods to carry out a detailed review on chronic diseases. The SCS Safe Contact Solution Analysis Project controls certain optimum-dependent methods for quality standards, timelines, relevance, specificity, affordability, utility, usability, confidentiality of the safety and confidentiality of decision-making. The compilation of data must be reliable and free from errors that are obtained from medical instruments and stored data in the back-end servers. The dimensional model for associating data models according to different categories, parameters, measurements metrics and standard benchmarks is used in a conceptual model of quality. Monitoring application gathers, processes and delivers data from the data centres. The quality of the indicator used to monitor, manage and develop the system is specified by data quality modelling.

Bhoomika B. K. et al. [3] identified the monitoring of healthcare device to assist the patient with routine control of the heart rate using four different components, such as the input of the patient, data contact, interface with the user and retrieval of results. A photoplethysmography sensor (PPG) is used to classify blood pressure and heart rhythm from time to time. Then using an analog device, a patient's pulse tracking can have separated converters for digital. Zigbee communication protocol, UART microcontroller, GSM and FT 232 IC modem are used for the communication of messages to physicians. Digital user interface is used to explain the state of the patient in the hospital while the doctor may be away from the patient, and the service can be done previously. If in case the heart rate goes down, the procedure model tests the patient's heart rate, the audiovisual signal produces a warning that it will be directed to all relevant doctors' unit rooms.

Raj Gaurav et al. [4] mentioned that the IoT-based EMG (electromyogram) detects pain and dislocation in nerves, muscles and tissues. It is possible to classify the discomfort of the worker, high load carried, nervousness of the worker (higher and lower) and weighing potential in a work environment. EMG sensors are equipped at the workstations to track any disruptions in the weighing of the cargo by people. Three main techniques are used: fast Fourier transform (FFT) is primarily used to transform the magnitude and frequency for rapid signalling; spectral signal transformation and time with a low rate of signal transformation are used for short-time Fourier transform (STFT) signalling; and time-frequency representation (TRF) uses time frequency for the simplified study of wave transformation for EMG signal. The ESP8266 Wi-Fi module is added to the conversion of the EMG device signal.

Chaudhury S. et al. [5] defines techniques that is based on IoT continuous monitoring of the health monitoring system on the internet, which uses a three-tiered architecture, mostly wireless body sensor networks (WBSN), base stations and GUIs in smart clinics. Data is accessible for humans and computers and mobile. Temperature sensor LM35, ECG AD8232 sensor and the GUI-connected heart rate monitor transforms messages from the GSM modem. An HTML page designed by the user will automatically upload information from the WBSN. Hospitals and telehealth services are effective and have customer support.

Jie W. et al. [6] defines wearable IoT used for health tracking assisted by computers, sensor network, data processing and APIs. RFID, 6Low and Zigbee are short-range communication linked with devices for disease-based diagnosis. EMG, ECG, EEG and blood pressure can be monitored using the wireless body area network (W-BAN). The biomedical system will continuously capture the data. In LCD-embedded systems, patient status recognition, tracking and monitoring, and regulating the system can be seen. Details can be submitted to wise database cloud servers.

Quan Zou et al. [7] describes quick, advanced medical health science–related machine learning techniques adopted by neural networks, decision trees and random forests. The two main components are taken into account in the experimentation of the dataset, especially principal components analysis (PCA) and minimal redundancy maximum relevance (mRMR) for the prediction of type 1 and type 2 diabetes. Machine learning is used for the assessment, variation and cross-validation of the dataset. Neural networks, pattern recognition and MAT laboratories used forward and feedback networks. For hospitals, Mohanty and Das introduced an IoT healthcare model to remotely track patients and connected equipment using a cloud-connected computer. The model incorporates a scope for reading barcode sensors that capture and process user authentication. The suggested device sensor is wireless and is used as part of an interconnected wireless sensor network within a specially built smart room to track babies.

Pradhan B. et al. [32] has discussed that IoT has shown potential growth in the health industry. IoT has helped various medical practitioners and health workers with the help of various medical devices and sensors. The author has also discussed various challenges and issues faced in healthcare with using IoT, which includes the high cost of maintenance, power consumption, scalability and security issues [33]. This research serves as a communication bridge inside the challenged community and the rest of the globe. The proposed work for Indian sign language recognition (ISLR) uses three-dimensional convolutional neural networks (3D-CNNs) and long short-term memory (LSTM) techniques for analysis. A conventional hand gesture recognition system involves identifying the hand and its location or orientation, extracting certain essential features, and applying an appropriate machine learning algorithm to recognize the completed action. The author has implemented the calling interface of the web application, WebRTC [34]. This paper analyses the quality of numerous unsupervised, supervised and other methods for the integrity and prediction of breast cancer. This research could provide various methodologies to better understand early cancer detection. Early detection for breast cancer can be a potential benefit in the management of this condition, not only does early treatment make it possible to heal it, but it also prevents its recurrence.

2.3 IOMT (INTERNET OF MEDICAL THINGS)

IoT is altering our existence such that we have never thought before. In comparison to the conventional model, everything in the IoT universe is called smart objects and are related to each other. IoT is characterized by interconnected network protocols, media and principles, as a complex self-configuring network of physical and virtual objects. These things have identities and features, are able to link to communication networks such as the internet, and are capable of connecting to sensors, retrieving data, networking and connecting. IoT should also be seen as a modern version of information and communication technologies (ICT).

In the healthcare market, IoT has also unlocked a fresh challenge, known as the Internet of Medical Things (IoMT), spanning from smart cities to smart appliances. By raising the quality of life and minimizing medical costs, IoMT provides valuable advantages for people's well-being. Wireless sensors that are used to automatically track the state of a patient's well-being and connectivity skills are vital components for delivering information to caregivers. Using the ability of current technology to provide patients with the right treatments and make their life safer are the first steps towards a smart healthcare environment. The other supporting technology that enables IoMT is artificial intelligence, which can help medical doctors in nearly any aspect of their proficiency, such as clinical supervisory. Artificial intelligence–supported IoMT systems can constantly monitor the health of people. Smart machines, smart homes and virtual assistants will deliver to the elderly and disabled persons the necessary support. Epidemic disorders can be monitored and stopped by combined data acquired from IoMT sensors and information received from the health system. Smart systems will allow authorities to provide the best assistance to civilians during crises and to take measures required in a given timeline. In the IoMT, input components, such as biosensors, networking modules and clients, work together in an effective and safe manner to deliver the best healthcare service. Via the help of IoMT technology, self-care and early detection, especially those using remote monitoring systems, are the major services in the healthcare ecosystem. The remote monitoring system is heavily dependent on the processing and analysis of real-time data gathered by biosensors. Safe mechanisms and communication technologies are necessary for data sharing between such devices. However, data loss and identity theft, if these systems are not sufficiently secured, are major issues [10].

2.3.1 IoMT applications

IoMT has opened new horizons in the health sector, together with other enabling technologies, to protect and enhance the quality of a person's life. It has developed new fields of applications and changed the way things are handled in traditional domains, like clinical decision-making, data gaining and maintenance of patient information.

With growths in the arena of AI, telemedicine and sensor-based technologies, IoMT can be made available for medical decision-making tasks. In hospitals, human judgements assisted by computer-generated assistants who have learned secret techniques from medical data using advanced algorithms, like deep learning, are now used. Through presenting fresh knowledge from numerous papers, journals and professional procedures for proper patient care, virtual assistance may keep doctors up-to-date. IoMT can be used to track patient life on an ongoing basis, can trigger warnings as health conditions change and can often take the appropriate precautions.

Blockchain is another supporting technology for IoMT. Using time stamps as data chunks linked to each other, a blockchain is described. The previous block's hash and data are used in a new block, thereby creating a chain. The finance industry has been revolutionized by blockchain technology, and we have seen numerous emerging innovations, like cryptocurrency. In the IoMT sense, blockchain technology has the ability to solve the incorporation, integrity, possession and access control of health data. It has been suggested for sensor data hash storage to avoid modifications and guarantee the ownership of patients. The supply chain will use blockchain technologies to use IoMT devices to monitor and track medications and medical products from the manufacturer to the end customer and avoid any errors. Sensors for temperature and humidity may be used during transportation and storage, for tracking environmental conditions. The management of medical information in a global ledger that all health data providers can access is another significant implementation area. Body-centric applications refer to healthcare systems or devices which, by directly communicating with the human body, produce psychological data. Acquired information is then stored and distributed to caregivers for analytical medicine. Health wearables and non-wearables are typical examples of devices that are body-centric. The intelligent sensing instruments that can be worn on a human body as implants or extensions are medical wearables. Non-wearables are smart sensor instruments that cannot be used worn over the bodies of people. Body-centric applications are split into two categories, indoor and outdoor, depending on the setting.

Object-centric applications may not specifically be linked to the person's body, unlike body-centric applications. Object-centric technologies refer to solutions and programs for healthcare that can be used effectively to improve the delivery of healthcare. Hospital information frameworks are a crucial example of object-centric implementations. Object-centric systems are again split into two zones, indoor and outdoor, on the basis of the environment [11].

2.3.2 IoMT components

There are three major components of the IoMT ecosystem: data collection, communication portal and server/cloud, as shown in Figure 2.2.

The primary aspect of IoMT, as it mainly relies on sensors and uses, is information acquisition. Patient data is captured through sensors and is passed through pre-processing phases prior to being stored on the server/cloud. In order to collect and interpret valuable information, various data collection methods are applied. This data is then utilized by medical professionals to be analysed. The information amid the components and user is

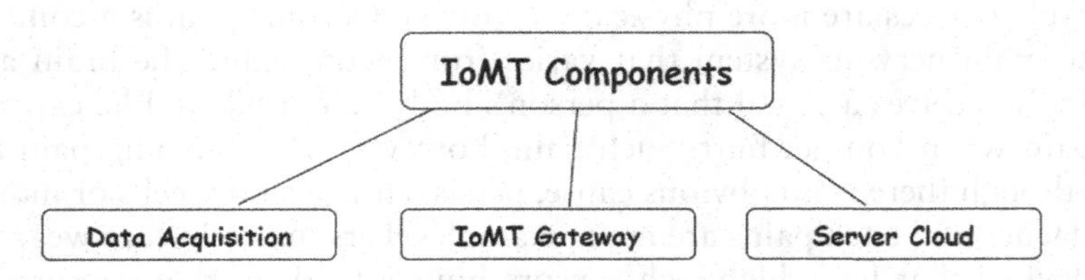

Figure 2.2 Main components of IoMT [9].

shared through gateways for communication. Both the three mechanisms of IoMT and their functionality are elucidated in this section.

In IoMT, information collection is for biological data procurement method for helpful purposes. Biosensors function in the form of an analog signal that captures the data. Biological signals are most commonly low in amplitude and polluted by noise. These signals are then pre-processed. Pre-processing involves operations for intensification. During data processing, it is crucial to guarantee that the data is persevered and is not missing; otherwise, it could contribute to incorrect choices with the diagnosis.

The electronic instruments utilized in biomedical signal attainment are biosensors. Through binding individuals to the health system, biosensors will radically change existing healthcare systems. This smart device can produce and send to caregivers an enormous amount of data. Often, by encouraging people to keep check on their health, biosensors can interpret the data automatically and accelerate diagnosis. Biosensors are classified as wearable and non-wearable technology. Wearable sensor is a type of intelligent detection instrument which may be in the form of implants or attachments which are worn on a human body. Such instruments provide high data collection and connectivity capacity with other machines without interference from humans. The following are some of the trendiest wearable technologies. Listening aids are tiny electronic instruments that are placed at the back of the ear: microphone, speaker and amplifier are part of a standard hearing aid system. Today, there are millions of persons suffering from the loss of hearing. Hearing aid systems allow people in silent and busy environments to hear clearly. The microphone absorbs and translates soundwaves from the atmosphere into electrical signals. An electronic amplifier strengthens electrical signals and then transfers them back to the ear via speakers. The oldest form of wearable technology is hearing aids, but now, they're much more advanced and intelligent. Rechargeable equipment that can be attached to mobile phones, televisions, PCs, computers and more are smart hearing aids. Users have the ability to change the volume and program of these machines using mobile apps according to their needs. Fitness trackers are similar to wristwatches; wrist-worn gadgets are fitness trackers or exercise bands that can monitor a person's physical habits along with other details, such as HR, distance travelled and temperature. Through Bluetooth, most exercise trackers can be attached to mobile phones. It is shown that consumers with wearable devices are more physically involved. Chronic pain is a condition of the main nervous system that varies from acute pain. The brain automatically receives a signal that a person's body is injured, and he can sense the pain when you get hurt. Such pain, however, is an ongoing pain that, even though there is no obvious cause, persists for several weeks or months. Headaches and back pains are main examples. Personal skin care wearables are devices that have high-tech sensors built into them that capture skin using Bluetooth technologies; information and data are transmitted to the smartphone app. A smartphone app analyses the data and, depending on your skin tone, selects the right LED color and skin patch. Patches are then applied with LED shining on the skin. Cancer cell–detecting wearables are used for cancer detection. Cancer is usually diagnosed after getting a sample from the body of the patient. This sample is then used to analyse cancer or tumor cells more closely in order to scan for them. In order to detect circulating tumor cells (CTCs) in the blood, taking a blood sample is another traditional procedure. Likewise, blood sketches may have detrimental health effects on cancer patients.

2.4 NON-WEARABLE SENSORS

A group of smart sensor instruments that cannot be equipped on a human body is non-wearable technology. Such devices can also produce large volumes of the data and connect with various devices. The following are some of the most trending non-wearable developments in biosensors:

2.4.1 Smart pill box

Owing to a demanding life, many individuals often combine their medicines. Non-timed medicine has many negative health effects. Patients must take medicine for a period of time, especially with chronic diseases. It is also imperative to take the proper dosage at the right time. Most of the time, elderly people are those who fail to take drugs or even overdose, which may have harmful effects.

2.4.2 Smart beds

These are used widely in hospitals. This system facilitates the remote control of patients by physicians or caregivers. In smart beds, biosensors for respiration, temperature and heartbeat are used. For sleeping measurement, heart rhythm and breathing rate habits of patients using smart bed technologies and long-term data may be checked. These devices can generate a warning in emergency scenarios or send a message to the caregiver for urgent action.

2.4.3 Human activity detection

Obesity, respiratory conditions, musculoskeletal illnesses and strokes have been the main health concerns. Health surveillance and human activity tracking devices provide the greatest potential for health rehabilitation advice and early warning for those problems. IoT technology helps one remotely run certain healthcare systems by the use of numerous wireless sensors and data collection.

The identification of human activity at home using numerous WSNs is a typical example. With such multiple interlinked sensors, however, high maintenance and implementation costs are expected, and high power usage is also required. Bodanese et al. are designing a low-power radar-enabled sensor network in order to track human activities. It examined 15 tasks conducted in the kitchen. The analysis found that 92.81 percent average activity detection accuracy is provided by the proposed method. In addition, it detects human activities more than 89 percent of the time in real-time detection mode [9].

2.5 PERSONALIZED MEDICINE USING IOT AND ML

PH has been broadly applicable to IoT and ML. In general, these applications are a series of integrated IoT-enabled devices for predictive research, diagnosis, remote control, preventive analysis and certain cases of surgery. Many technology firms have invested in the development of a personal healthcare infrastructure focused on machine learning. The researchers have developed algorithms in the academy to teach the computer to identify multifaceted indications from high-dimensional images. There are, however, several difficulties and issues in the use of IoT and ML for PH.

2.5.1 Personalized healthcare

The current medical pattern is that physicians most commonly recommend medications through the testing and trying approach. And for each person, the reaction to a drug is not the same. Although most persons will respond to a given prescription and dosage, the same treatment may have either mild or serious side effects in a subset of patients. PH will increase the superiority of treatment and lower costs at the same instance. It can even support us in predicting the best treatment with the fewer contrary effects for individual patient.

A promising case study of PH with IoT and ML is on personalized diabetic treatment. A person's eating patterns and insulin reactions are different. Often, the body's reaction to food consumption ranges from human to human. Bluetooth-activated blood sugar level tracking provides the patient with nutritional advice about at what time and what quantity of the food can be consumed, and also with dietary advice on the effects of feeding without supervision. You will get this advice from your own mobile device.

In cases of patient readmission, PH still plays a critical function. Post-surgery patients were readmitted to the hospital, according to a fifth report. Of these, if a good customized healthcare package and surveillance are in effect, one-third of the readmission might have been stopped. Researchers found that insufficient coordination of treatment was responsible for the unsustainable spending of billions on preventable complications and excessive patient readmission in local hospitals in the United States. The principle of applying IoT and the Internet of Nano Things (IoNT) was the progression of lower-cost, smaller calculating devices. In addition, the advantage of these innovations is understood in the application of healthcare.

ML and IoT can be used in the mentioned areas, as shown in Figure 2.3. The areas are as follows:

1. Diagnostic care
2. Assistive care
3. Monitoring and alarm

A biomodule device can help do certain diagnosis of the patient in diagnostic treatment. For instance, professional doctors are not available in remote areas. There is also a shortage of pathological professionals and operators of surgical equipment in regions which are rural and semi-remote. These problems can be solved with intelligent and automatic diagnostics. Diagnostic facilities with IoT and ML will also advance the PH tremendously.

Assistive PH programs focused on ML and IoT have also had a consequence on many people's lives. The effect is rising because of technical change. Assistive PH, though, would still require to fix accessibility, technological hype vs. actual value and affordability challenges. In addition, the problem of safety and security for small devices will create considerable concern, as they can be compromised and altered if not adequately protected.

Inside the PH service, remote control plays an important role. As part of PH, home surveillance will be used for patients relieved from the hospital, people living remotely and elderly people living in residential care. The personal health records of patients can be moved into the cloud system through the remote control of the health status by IoT devices. In addition, big data services focused on the cloud can work on smart statistical analysis based on the PH data stored [12].

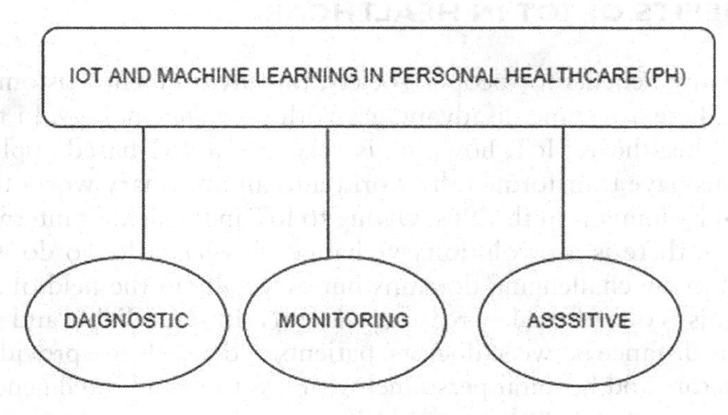

Figure 2.3 IoT and machine learning roles in personal healthcare [12].

2.6 APPLICATION AND CHALLENGES OF IOT IN HEALTHCARE

All the services we analysed are from the medical healthcare networks. Many of the demands came from academic papers released in the year 2016. So our applications that have been researched are up-to-date. These applications were chosen on the basis of the contributions of various IoT researchers from different resources. We have chosen papers from numerous authentic repositories, such as ACM, IEEE and Elsevier, based on a collection and rejection criteria. We also included the most recent applications with respect to healthcare in the field of IoT. The aim was to list IoT applications within the area of medical healthcare.

2.6.1 Challenges of IoT in healthcare

We listed some major challenges in the IoT domain after a brief analysis. We assume that we will increase the IoT standard in the field of medical care if these problems are faced in the area of IoT. In the field of medical care, IoT will provide more efficient and quality services. There is a disruptive shift in the field of internet communication due to IoT; IoT contributes to the advancement of many difficult fields but particularly in the field of medical devices. This is one of the key reasons why it is simple, reliable and scalable to close the distance between doctors, patients and healthcare providers. IoT allows doctors and hospital personnel with less time and intelligence to do their job more reliably and actively. Challenges were chosen on the basis of the contributions of various IoT researchers from different resources. We have chosen papers from numerous authentic repositories, such as ACM, IEEE and Elsevier, based on a collection and rejection criteria.

2.7 BENEFITS OF IOT IN HEALTHCARE

IoT has many benefits for people, society, the environment, customers and business. There are some disadvantages with every technology. In the field of medical healthcare, IoT, however, is very useful. IoT-based applications and systems have transformed the world into an imaginary world that was thought of by humans in the '90s. Owing to IoT in the field of internet communication, there is a revolutionary change; this has a lot to do with the growth of many challenging domains but especially in the field of medical devices. This is one of the key reasons why it is simple, reliable and scalable to close the distance between doctors, patients and healthcare providers. IoT allows doctors and hospital personnel with less time and intelligence to do their job more reliably and actively [13].

2.7.1 IoT for doctors

In sectors such as medicine and dentistry, the use of the Internet of Things (IoT) is increasingly growing to produce evidence-based results. This technology is now embedded into orthopedics to provide appropriate knowledge for treatment one day. In order to increase the consistency of care, IoT utilizes sensors. These sensors easily provide the measurement of bone and fracture data. This technology reliably offers knowledge about blood pressure, pain in the bone and operation of the brain. It guides patients after an accident to exercise properly for quick healing. It increases customer loyalty, productivity and performance successfully. In the field of orthopedics, IoT promises improvements in treatment, surgery, knowledge, schooling, growth and science. After the surgery in orthopedics, successful coordination between orthopedics and patients, and with its help, there are better chances of recovery for the patient. It collects the patient's exercise and daily routine data and alerts them to follow-up after surgery for recovery. Various software applications are used to capture patients' knee and hip replacement data. This data is easily shared with doctors for the patient to recover.

Orthopedic surgery requires a great deal of physical exercise and in-home therapy. This technology monitors the patient without any need for healthcare professionals. Surgeons and patient care teams send a message to remind them and guide them step-by-step to improve patient outcomes through better exercise. Without any doctor's appointment, patients and their families can quickly monitor recovery progress. It solves the problems in question during the patient's confusion. Reducing operational performance is the main advantage of this technology. During surgery, there are costs and errors. It handles all drugs properly; this virtual infrastructure will improve patient outcomes. All physical orthopedic devices are connected to the internet by IoT receiving or sending information quickly. This digital health medical devices can easily send data to the record system, physicians, surgeons, doctors, laboratories and related organizations. This innovative technology has an outstanding ability to achieve quality outcomes in orthopedics with less physician involvement. During surgery, in order to create excellent service, this carries out better treatment of complex cases. It helps to achieve precision, dependency, reliability and adequate treatment medication.

Various technologies are used in IoT. It is helpful in orthopedics to provide adequate information. Big data, cloud computing, smart sensors, artificial intelligence, actuators and augmented reality/virtual reality are included. The process used in IoT for orthopedics is shown in Figure 2.4.

Different IoT technologies are used to store data on a server, and the associated information required is collected. This information is the best solution for checking, updating and alerting patients about future diseases and their sources, and for preventing inadequate medication and treatment. It reduces

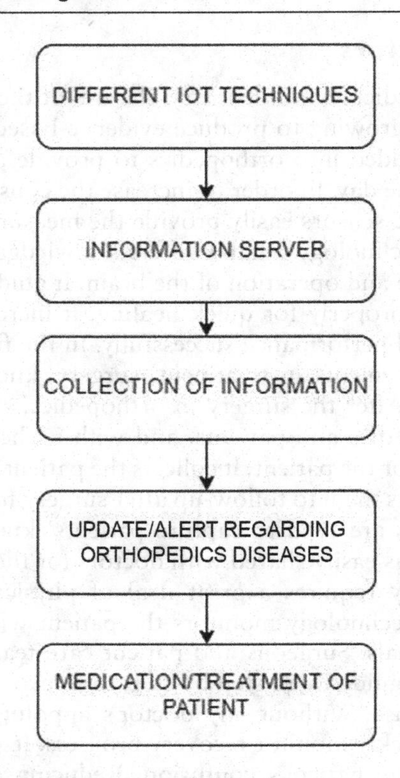

Figure 2.4 Process chart of IoT in orthopedics [14].

errors and detects errors during orthopedic surgery with the help of proper instructions and information. In the future, all history of illness, complaints and progress will be recorded and accessed for better therapy. To provide the proper treatment of a critically ill patient, it continually monitors the system.

In collaborative preparation and interaction with orthopedic surgeons, IoT is currently used. This created a big influence on training of nursing and clinical performance. IoT support gathers all orthopedic surgery data which allows doctors to make surgery productive. It stressed that an experienced surgeon must appreciate the goal of surgical procedures. There has been a huge change in research, preparation and orthopedic surgery with the introduction of this technology [14].

2.7.2 IoT for patients

Modern e-health systems focused on the usage of knowledge and communication technology will help to avoid health system issues and turn the health system into preventive, that is, remote surveillance of patients will help us identify improvements in their disease pattern and generate alerts accordingly for use by nurses or medical person before approaching a critical condition actively. It is realistic to have complete control over the condition of patients somewhere at all times, or in other words, both healthcare and social ecosystems (patients, social workers, family members and doctors) will be benefited.

With the help of IoT, the patient's health can be monitored through the wearables and other body sensors. With IoT, the patient's health can be monitored without having the patient admitted to the hospital. His health can be regularly monitored from home. IoT provides accurate and real-time information about the patient [15].

Chronic patients and dependents use a smart mobile system that receives data from vital signs sensors and transfers this data through the 3G/HDSPA mobile network or the internet. The architecture also enables a smartphone to be used as a devoted mobile computer for processing intermediary data. A method of rule generation for warning activation can be submitted to the caring staff or nursing or medical personnel of the chronic patient. In reality, this is the centre of the framework for producing innovative e-health services and early detection. Interoperability and knowledge delivery mechanism for all participants of the scheme leverages the new technical developments in communication. A website portal can let both social services and family caregivers refer to the relevant patient details from both a personal computer and from mobile devices. A health-based knowledge-seeking module, in this situation, may be a web interface where the user can search for medical terminology related to their health status using natural language; this module gathers data from different sources, such as MedlinePlus or Free Base [15].

2.7.3 IoT for hospitals

The introduction of smart devices to hospital will save costs in operations and increases patients' medical expertise. The labor rate of medical workers is reduced. Narrowband IoT (NB-IoT) is a wireless wide-field protocol that operates nearly everywhere with low power. NB-IoT has the benefits of low cost and low energy consumption, which allows a new way to connect devices that need small volumes of data in difficult-to-access areas for long periods and has been used in smart devices, smart meter reading, parking and bicycle sharing. The emergence of NB-IoT allows the formalization of architecture for linking all smart things in intelligent hospitals. Different sensors inside or outside the hospital gather information, including reports on physiological indicators, such as heart rate and blood pressure; environmental statistics, such as temperature; and other data, such as data from the parking room. Such information is ultimately sent to the cloud network for decision-making and analysis through wireless communication. Figure 2.5 depicts how IoT can be used in various scenarios in hospitals.

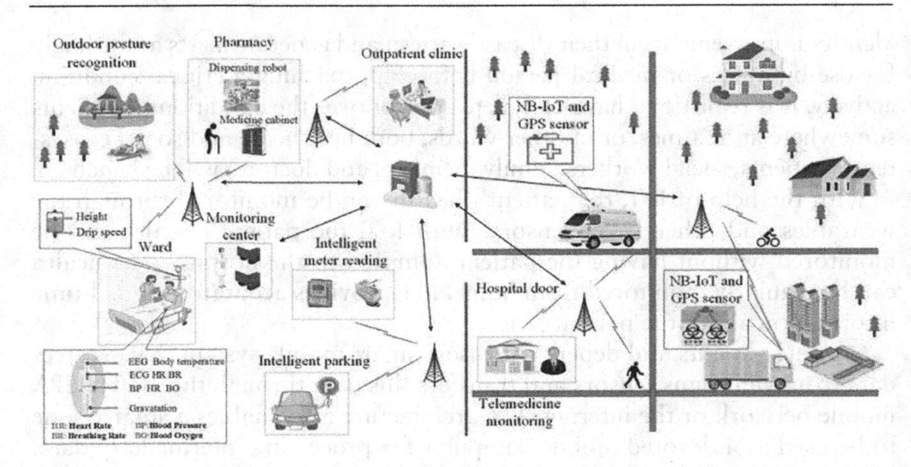

Figure 2.5 Illustration of the application scenarios for smart hospitals [16].

2.7.4 Application scenario

IoT helps in intelligent parking. Parking space can be managed by intelligent locks in hospitals. Before he goes to the hospital, a patient can book a parking spot using a smartphone app. When the patient enters and given an unlock order via wireless contact, the reserved parking spot will be closed. The parking spot can be auto-locked as the patient exits the hospital, and the price settlement can be done automatically. Any essential hospital branch requires access control devices to be installed. Using wearable sensors, commands may be sent to the machine to complete the authentication when a worker enters a building entrance. The door will open if the authentication is effective. Several employees can also perform remote unlocking if needed. IoT can also help in ward care. In the ward, wearable monitors or smart sensors will capture real-time physiological indications of the patient, like heart rate, or environmental details, such as cleanliness. Such data is then transmitted by wireless communication to the monitoring centre. If the medical sign of the patient is irregular, the paramedics will do the corresponding procedure in time. IoT can be used for outpatient medical treatment. Outpatient clinicians may develop a thorough knowledge of the health of the patient based on wearable physiological sign information gathered. Apps that can help physicians make correct diagnosis increase the efficacy of diagnosis by the doctor and save time for the patient. Outdoor posture recognition: whenever the patient is outside the gate, posture sensors will detect the body and movement posture of the patient to assess whether a risky posture has been generated; that would also evaluate whether the inappropriate physiological indication is a false warning for normal circumstance. Telemedicine monitoring: at home, patients continue to be tracked. Wearable computers are capable of remotely monitor the patient's vital symptoms. If the physical state of the patient is irregular, the device informs the families of the patient or the attending doctor on time to prevent accidental occurrences. Also, intellectual meter reading can be accomplished by connecting a wireless communication unit to the hospital for conventional power and water meters. It is possible to attach any costly medical equipment such as gamma rays to the IoT device, so the equipment testing can be done occasionally. If a valuable medical item or an important medical item that is medical waste is eliminated, the sensors mounted will deliver the place and status data in real time to the cloud portal for efficient tracking [16].

2.8 APPLICATION OF MACHINE LEARNING IN THE HEALTHCARE INDUSTRY

Data science and ML have shown accurate results in the healthcare industry. ML is helping a lot in patient care strategies. Most common techniques used in healthcare are deep learning techniques, like computer vision, reinforcement learning and natural language processing.

A few benefits of ML in healthcare are as follows:

- Accurate and efficient medical diagnosis
- Reduction in the load of administration health professionals
- Prevention in delay of reporting emergency cases
- Enhancement of diagnosis speed

Machine learning application in healthcare are as follows:

Medical imaging: Image recognition is used in computed tomography and magnetic resonance processes for detection and prediction of diseases. Deep learning (DL) models make practical interpretations, like tissue shape, volume and size, with imaging data. Also, these models are used for the initial detection of diabetic nephropathy and for ultrasound to detect breast nodules and Alzheimer's disease.

Deep learning algorithms determine abnormalities by simplifying complex data. The visions provided by CNNs help health professionals diagnose well the problems of patients on time.

Mental chatbots: Happify and Wysa are AI-based mental health applications that provide realistic conversations with patients to know more about their problems and issues.

Patient health and analysis: Deep learning models can learn from unstructured and structured data that includes clinical notes and laboratory tests. These models diagnose at faster speed and with more accuracy.

Errors in prescription: Deep learning (DL) models can notice the errors in prescriptions as well. These models audit the prescriptions and patient records to find and detect errors in prescription and to correct those errors.

Fraud detection for medical insurance companies: Deep learning models can help insurance companies to detect such fraudulent cases. Deep learning models detect and analyse healthcare data and fraudulent behaviors, thus, helping the insurance companies.

Machine learning may be shown to recognize and help detect patients who may be more vulnerable to recurring diseases. In comparison, almost 90 percent of trips to emergency departments are preventable. To help classify and direct patients to proper treatment, machine learning may be used, while reducing costs by discouraging patients from expensive, time-intensive treatment centres [17].

2.8.1 Drugs discovery

The past discovery of drugs is evidence to the growing promise of computer-generated exploration of results. The only factor in drug discovery is the projected cost of producing medications raises every few years. In this dynamic climate, another approach to potentially increase efficiency is to apply ML algorithms in all fields of drug research and development of adequate evidence to train models. ML is an increasing artificial intelligence field that uses various mathematical techniques by which computers can learn from different types of data without being specifically programmed. The creation of nano-medicines can extend to machine learning by utilizing massive datasets similar to other fields in finding medicines. These activities will make it possible to quantitatively model suitable molecules prior to synthesis and concentrate research of the most promising applicants on experiments. The domain of nano-medicine has contributed to the growth of nano informatics and the use of data mining and ML to produce nano-QSARs to forecast the functional and structural properties of nanoparticles. A surprisingly large variety of approaches to machine learning was applied to the estimation of various nanoparticle biomedical properties, such as the prediction of molecular release, nano-loading, cellular uptake, cytotoxicity, molecular uptake, loading, adherence to nanoparticles, molecular release, polydispersity and size. In order to predict, the particle self-assembly mechanism for targeted nanoparticles of drug carriers, computational methods can be used. Nanoparticle assembly prediction models with quantitative structure were used to produce nano-assembly predictions that were found to capture drugs with heavy loading and were then also examined in cancer models. In order to classify clinical experiments involving nanodrugs and nanodevices from clinical trials, a method of machine learning has also been identified. While machine learning implementations have seen decades of drug discovery, there are few examples of nanoparticle research, and data available for model creation is limited to selecting databases of nanomaterials. Since minor datasets are used in most of the published examples, deep learning has minimal utility and has seldom been implemented. To date, commercially accessible instruments that could enable scientists to create nano-medicines are yet to be developed using these models. The importance of methods relevant to nanomaterials is in the formulation or distribution of drugs and should be considered an integral part of E2E [18].

2.8.2 Cancer diagnosis and prognosis

Cancer has been characterized as a heterogeneous disorder composed of several distinct subtypes. The early detection and prognosis of a form of cancer has become a priority, as it will promote future clinical treatment of patients. The value of classifying patients with cancer as being in high- or low-risk categories has prompted numerous biomedical and bioinformatics research teams to explore the use of machine learning (ML) methodologies. Consequently, these approaches have been used to model the development and treatment of cancer disorders. Moreover, the ability of ML software to detect key features from complicated datasets indicates their significance.

A number of various ML methods and features have been used over the past two decades. Selection algorithm has been widely used in cancer prognosis and disease prediction. Most of these works use ML strategies to model cancer progression and define insightful variables and are included in a classification system afterwards. Additionally, in nearly all gene expression profiles from research, as clinical factors, in comparison, historical parameters are used in a complementary way in order to be fed to the prognostic protocol as input: (i) susceptibility to cancer, (ii) recurrence and (iii) ML methods of survival were used for prediction. The data was obtained based on a number of query searches in the Scopus biomedical database. In fact, the number of papers resulted in inquiries such as "cancer risk assessment" and "machine learning", "cancer recurrence" and "machine learning", "cancer survival" and "machine learning", as well as "cancer prediction" and "machine learning". The effectiveness of a disease's prognosis relies, without a doubt, on the accuracy of a medical diagnosis; moreover, there should be a prognostic forecast, taking more than a basic diagnostic judgement into consideration. One is concerned with three statistical activities when dealing with cancer prediction: (i) estimation of cancer sensitivity, (ii) prediction of cancer reappearance and (iii) cancer survival estimation. The other seeks to discover (i) the risk of contracting a type of cancer in the first two cases and (ii) the probability of redeveloping a type of cancer following complete or partial remission. Lastly, the key goal is to predict the results of recovery, such as disease specifics or total survival following cancer diagnosis or treatment. The cancer outcome calculation generally refers to cases of (i) life expectancy, (ii) survival, (iii) growth, (iii) progression and (iv) response to treatment.

The main categories of ML techniques used in cancer diagnosis for decades have been ANNs and decision trees. One or more ML algorithms are used by the vast majority of these publications and combine data from heterogenous sources for the identification of tumors and for cancer type diagnosis and prediction. An increasing trend in the use of other supervised learning methods, including SVM and BHs, in cancer detection and prognosis have been noted in the last decade. Many of these algorithms for classification have been widely used in cancer study issues [19].

2.8.3 Medical imaging

Advancement in both imaging and machines has made rapid growth in the possible application of AI in various tasks for radiological imaging, like measurement of the risk, prediction, analysis, prognosis and reaction to treatment, including the exploration of multi-omics diseases. The primary undertaking of radiologists is medical image analysis, with the duties involving good visual consistency as well as good understanding of images. Human image perception is constrained by the existence of structural noise, incomplete visual search patterns, exhaustion, disturbances, examination of visual search patterns, distractions, subtle or dynamic states of illness, large quantities of data from images and the image's physical consistency itself. Successful disease diagnosis and treatment rely on the integration of biological, molecular, imaging and genomic evidence (i.e., separate omics) information from various patient studies. An extension of CADx, radiomics, has been defined as the translation of images into minable data. Segmentation of a tumor by device from its history accompanied by computer extraction of different tumor characteristics can require obtaining radiomic data. Quantitative radiomics has the ultimate advantage of generating predictive image-based disease phenotypes for precision medicine and generating phenotypes of image-based data mining for other phenotypes of data mining with -omics to scan for (i.e., imaging genomics). Radiomic characteristics can be classified as manufactured or engineered, or deep-learned features or intuitive features. Cancer is a significant subject of radiography. Cancers are spatially varied, and cancer biomarkers typically only include measures of size and fundamental enhancement of cancer tumors (if dynamic imaging is used). The heterogeneity in primary breast cancer tumors has been shown by multiple genomic studies. The concern is to obtain image-based phenotypes with radiomics of the cancer containing tumor, with size, form, margin morphology, enhancement texture, kinetics and variance kinetic phenotypes. For example, in the first postcontrast, the improvement of texture phenotypes characterize the tumor texture pattern in the photographs of contrast-enhanced tumors and, thus, quantitatively characterize the heterogeneous nature of contrast absorption within the breast tumor. Radiomics also enables computer-extracted lesion characteristics to be used as image-based phenotypes to predict the reaction of a patient to a given clinical procedure. The functional volume of the tumor, for instance, from breast MRI in an assessment using data from an ACRIN analysis, can be an indicator of survival in patients undergoing neoadjuvant therapy. A subcategory of ML in deep learning is the use of multi-layered networks to evaluate dynamic patterns in raw imaging input data [20].

2.8.4 Smart health records

A body sensor network (BSN) is a modified network intended to function alone internally and externally of the human body to numerous medical sensors and implants. Its integration in medical supervision would give both healthcare practitioners and patients versatility in service and cost saving choices. They decrease the irritation of users and improve mobility. In applications, monitoring of human physiology in this group requires patient info, records and reports in a hospital, hospital drug management and so on. Vital signs are used to indicate whether to calculate the basic functions of the body that can be useful for monitoring a person's general well-being. It is really important to provide healthcare facilities for those who have chronic illnesses in particular. These people require continuous treatment that cannot be administered outside the hospital. Figure 2.6 depicts how various sensors can be used in health monitoring.

The fundamentals of the Arduino board, various sensors and LabVIEW software are covered in this section.

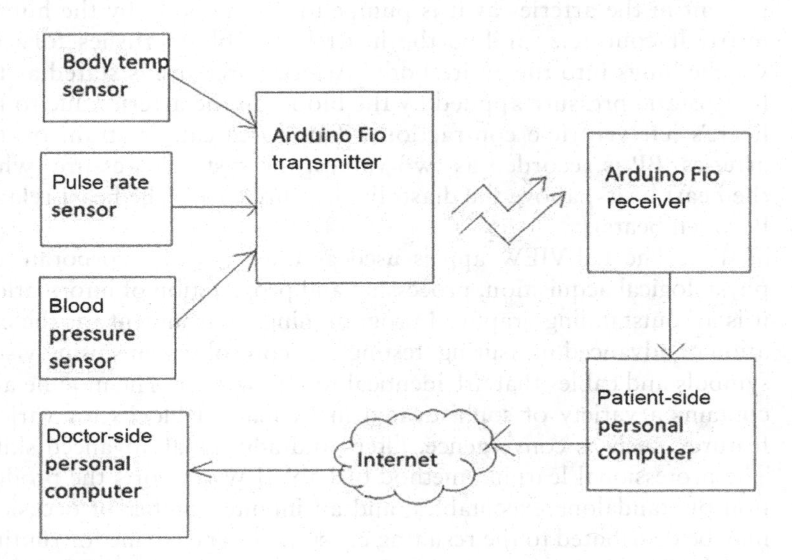

Figure 2.6 Representation of the system [21].

2.8.4.1 Arduino fio

The Arduino Fio–based Zigbee Development Platform is an 8-bit development platform that provides the means to convert code. It is based on the ATmega328P AVR microcontroller and has knowledge of all the 8-bit microcontrollers. The Arduino Fio is for software that are wireless.

The user should use a USB cable to import the sketches (program) or use a tweaked adaptor for USB-to-XBee, such as an Arduino Fio transmitter (base). The software may be imported by the user using XBee S1 wireless. A variety of projects can be built using the framework in different fields, such as wireless networking, robots and consumer electronics:

Temperature measurement: Using an IC LM35, the temperature sensing can be done. The LM35 is an integrated circuit temperature accuracy monitor with a linear output voltage proportional to Celsius or Centigrade temperature.

Pulse rate measurements: The IR sensor pair essentially consists of an IR LED and a photodiode, which is generally referred to as a photo coupler. The IR sensor works on the assumption that IR LEDs emit IR radiation, and the photodiode detects the IR radiation. The resistance of the photodiode will change due to the amount of IR radiation that falls on it, and thus, the voltage decrease often varies through it, and we can feel the voltage comparator (LM358) by using it, changing the voltage and producing the output.

Blood pressure measurements: Blood pressure is the pressure of blood present in the arteries as it is pumped into the body by the human heart. It contracts, and as the heart beats, blood rushes forward via the lungs into the entire body. Arterial pressure is stated as the hydrostatic pressure applied by the blood on the arteries due to the heart's left ventricle contractions. This force causes strain on the arteries. BP is recorded as two values: the systolic pressure (when the heart beats) above the diastolic pressure (when the heart relaxes between beats).

LabVIEW: The LabVIEW app is used as the basis to incorporate the physiological acquisition, processing and propagation of information; it is an outstanding graphical programming environment for the creation of advanced measuring, testing and control systems using visual symbols and cables that are identical to a flowchart. The module also contains a variety of sophisticated mathematical blocks for various features, such as convergence, filters and additional advanced skills. The professional learning method of LabVIEW supports the production of standalone executables, and an infinite number of occasions may be distributed to the resulting executable. The engine for runtime and the libraries and executables can also be freely supported [21].

2.9 DISADVANTAGES AND CHALLENGES OF IOT AND ML IN HEALTHCARE

The electronic personalized healthcare sector is not prone to scrutiny and disadvantages. This inherits the latest simple fundamentals IoT and ML problems. A situation when an old person uses a sensor-based PH system is described subsequently. The sensor captures numerous data and transmits it to a database, such as heart rate, EEG, blood sugar and pressure. There, the data is stored for use by the parties concerned. The database uses some machine learning at the same time and algorithm for the study of the data gathered to determine the danger factor of the patient, enhance fitness and recommend more actions which are based on that. The disadvantages of transmission and packet decline; privacy and sensory interface authentication problems need to be targeted.

In general, the efficiency of the technology will be calculated by its usefulness to the customer. BAN uses sensors for body collection for the users' linked health records. BAN must also efficiently transmit and turn sensed phenomena into useful information in order to provide the PH service, and it must ensure that it satisfies other device specifications, such as quality of electricity. In addition, it must have the power to store selectively and distribute information at levels of correctness and speeds that are appropriate to the data destination; for example, if it's an athlete who is worried with her heart rhythm or a doctor who needs a patient to get an electrocardiogram. These different implementation criteria require the ability to collect hierarchical information and to implement BAN structures in the current architecture of IT.

ML has a close association with mathematics, decision-making from current research and prediction from the past experiences. The ML-based approach would evaluate the condition according to a qualified dataset in the case of tracking a patient. Most of the time, the training dataset plays a key role in effectively forecasting the future pattern of a given new problem. This dataset may sometimes be skewed to include several examples and might not be diverse. Noisy data, dirty data and missing data may contribute to the identification and estimation of a health-related diagnosis and advisory alert with a lower likelihood of ration. Sleep habits and sleeping schedule can vary from person to person, age and health status in the event of sleep observation and sleep apnea tracking. Therefore, a comprehensive dataset of all case studies will not be available to track sleep patterns, and this may result in the wrong PH estimate.

When IoT and ML are used to activate PH, the device will need to consider whether to diagnose, forecast and warn the user. There are several situations in which an ML-based decision may be inaccurate, and it is difficult, almost impossible, to figure out that a certain decision has been made.

Using the ML-based PH service, predictive analysis will help hospital-released patients who will need to be readmitted to the hospital build a risk stratification model where those patients with higher risk are handled with extra effort; predictive analysis is convenient. Such an undertaking requires, but is not limited to, the procurement of new surveillance facilities and continuous monitoring. In certain instances, based on statements and historical evidence, certain models are developed. The dynamic PH model that will help readmission prevention efforts would also exploit and use dynamic patient data to forecast possible scenarios and launch an action plan to decrease possible complications. For this case, the PH scheme may be used. However, the problems of the lack of data transfer, data noise and missing data need to be discussed [12].

2.10 CONCLUSION AND FUTURE SCOPE

People's dedication and understanding of their health are growing substantially. Owing to this, the need for home or remote monitoring is also proliferating. Researchers are investing a great deal of money and time into developing technical innovations and incorporating current healthcare improvement technologies. This paper summarizes IoT and ML interdisciplinary dimensions, including the categorization of smart healthcare systems, important sensing devices and connectivity modules. The primary purpose of this analysis is to provide an understanding of recent research attempts to improve and support the IoT ecosystem. The IoT elements of both software and hardware were thoroughly explored and explained. In addition to their core features, software modules, such as signal improvement and machine leaning, were also introduced. The following are the conclusions identified after the previously mentioned technologies and problems have been reviewed. The data is obtained through the use of sensory, wearable and non-wearable devices with sensors. Sensor devices also minimize the workload of caregivers and help to reduce the cost of healthcare. Therefore, sensory innovations should be viewed as part of interdisciplinary study in the healthcare sense since they potentially provide additional knowledge in healthcare information systems.

In addition, exchanging knowledge between sensors and healthcare systems requires suitable technologies for communication. By changing the hospital-centric world to a patient-centric world, the large wave of AI-driven IoT-based innovation will expand the boundaries of healthcare outside hospital settings.

In the future, IoT devices can track the health of patients in real time in medical emergencies like asthma attacks and heart failures. The collected data is transmitted in real time to the physicians, and a reduction in readmission rates of patients can be achieved. Here, unauthorized access to the patient's data may lead to serious security issues; hence, some research on making the data only available to authorized users should be carried out so it will improve data security.

REFERENCES

[1] Farahani, B., Barzegari, M., Aliee, F. S., & Shaik, K. A. (2020). Towards collaborative intelligent IoT eHealth: From device to fog, and cloud. *Microprocessors and Microsystems*, *72*, 102938.

[2] Claudia, C., & Gutiérrez, R. M. (2011). e-Health monitoring applications: What about data quality. In *HamIS 2011 Health Ambient Information Systems Workshop 2011*. Gramado, Brazil.

[3] Godi, B., Viswanadham, S., Muttipati, A. S., Samantray, O. P., & Gadiraju, S. R. (2020, March). E-healthcare monitoring system using IoT with machine learning approaches. In *2020 International Conference on Computer Science, Engineering and Applications (ICCSEA)* (pp. 1–5). IEEE.

[4] Gaurav Raj, N. R. (2017, July). IoT based EMG monitoring system. *International Research Journal of Engineering and Technology (IRJET)*, *4*(7).

[5] Saha, D. P. (2017). Internet of Thing based healthcare monitoring system. In *IEEE Annual Information Technology, Electronics and Mobile Communication Conference (IEMCON)* (pp. 531–535). Vancouver, BC. doi:10.1109/IEMCON.2017.8117245

[6] Wan, J., AAH Al-awlaqi, M., Li, M., O'Grady, M., Gu, X., Wang, J., & Cao, N. (2018). Wearable IoT enabled real-time health monitoring system. *EURASIP Journal on Wireless Communications and Networking*, *2018*(1), 1–10.

[7] Deepti Sisodia, D. S. (2018). Prediction of diabetes using classification algorithms. *Procedia Computer Science*, *132*, 1578–1585. doi:10.1016/j.procs.2018.05.122

[8] Joyia, G. J., Liaqat, R. M., Farooq, A., & Rehman, S. (2017). Internet of medical things (IOMT): Applications, benefits and future challenges in healthcare domain. *Journal of Communication*, *12*(4), 240–247.

[9] Al-Turjman, F., Nawaz, M. H., & Ulusar, U. D. (2020). Intelligence in the Internet of medical things era: A systematic review of current and future trends. *Computer Communications*, *150*, 644–660.

[10] Zikria, Y. B., Kim, S. W., Hahm, O., Afzal, M. K., & Aalsalem, M. Y. (2019). Internet of Things (IoT) operating systems management: Opportunities, challenges, and solution. *Sensors*, *19*(8), 1793.

[11] HealthSense: A medical use case of Internet of Things and blockchain - IEEE Conference Publication (n.d.). https://ieeexplore.ieee.org/document/8389459 (accessed November 7, 2019).

[12] Ahamed, F., & Farid, F. (2018, December). Applying Internet of Things and machine-learning for personalized healthcare: Issues and challenges. In *2018 International Conference on Machine Learning and Data Engineering (iCMLDE)* (pp. 19–21). IEEE.

[13] Joyia, G. J., Liaqat, R. M., Farooq, A., & Rehman, S. (2017). Internet of Medical Things (IOMT): Applications, benefits and future challenges in healthcare domain. *Journal of Communication*, *12*(4), 240–247.

[14] Haleem, A., Javaid, M., & Khan, I. H. (2020). Internet of things (IoT) applications in orthopaedics. *Journal of Clinical Orthopaedics & Trauma*, *11*, S105–S106.

[15] Páez, D. G., Aparicio, F., de Buenaga, M., & Ascanio, J. R. (2014, December). Big data and IoT for chronic patients monitoring. In *International Conference*

on Ubiquitous Computing and Ambient Intelligence (pp. 416–423). Cham: Springer.

[16] Zhang, H., Li, J., Wen, B., Xun, Y., & Liu, J. (2018). Connecting intelligent things in smart hospitals using NB-IoT. *IEEE Internet of Things Journal*, *5*(3), 1550–1560.

[17] Bhardwaj, R., Nambiar, A. R., & Dutta, D. (2017). A study of machine learning in healthcare. In *2017 IEEE 41st Annual Computer Software and Applications Conference (COMPSAC)* (vol. 2, pp. 236–241). IEEE.

[18] Ekins, S., Puhl, A. C., Zorn, K. M., Lane, T. R., Russo, D. P., Klein, J. J., . . . & Clark, A. M. (2019). Exploiting machine learning for end-to-end drug discovery and development. *Nature Materials*, *18*(5), 435.

[19] Kourou, K., Exarchos, T. P., Exarchos, K. P., Karamouzis, M. V., & Fotiadis, D. I. (2015). Machine learning applications in cancer prognosis and prediction. *Computational and Structural Biotechnology Journal*, *13*, 8–17.

[20] Giger, M. L. (2018). Machine learning in medical imaging. *Journal of the American College of Radiology*, *15*(3), 512–520.

[21] Vippalapalli, V., & Ananthula, S. (2016, October). Internet of things (IoT) based smart health care system. In *2016 International Conference on Signal Processing, Communication, Power and Embedded System (SCOPES)* (pp. 1229–1233). IEEE.

[22] Loureiro, T. C., Neto, A. B., Rocha, F. A., Aguiar, F. A., & Fernandez, M. P. (2019, June). Multi-agent system and classification algorithms applied for ehealth in order to support the referral of post-operative patients. In *International Symposium on Ambient Intelligence* (pp. 11–18). Springer, Cham.

[23] Snezana, S., et al. (2020). Cloud based personal health records data exchange in the age of IoT: The cross4all project. In Dimitrova, V., Dimitrovski, I. (eds.) *ICT Innovations 2020. Machine Learning and Applications. ICT Innovations 2020. Communications in Computer and Information Science*, vol. 1316. Cham: Springer. doi:10.1007/978-3-030-62098-1_3

[24] Pap, I. A., Oniga, S., & Alexan, A. (2020, May). Machine Learning EEG Data Analysis For eHealth IoT System. In *2020 IEEE International Conference on Automation, Quality and Testing, Robotics (AQTR)* (pp. 1–4). IEEE.

[25] Godi, B., Viswanadham, S., Muttipati, A. S., & Samantray, O. P. (2020, March). E-Healthcare Monitoring System using IoT with Machine Learning Approaches. In *2020 International Conference on Computer Science, Engineering and Applications (ICCSEA)* (pp. 1–5). IEEE.

[26] Pandey, H., & Prabha, S. (2020). Smart health monitoring system using IOT and machine learning techniques. In *2020 Sixth International Conference on Bio Signals, Images, and Instrumentation (ICBSII)* (pp. 1–4). doi:10.1109/ICBSII49132.2020.9167660.

[27] Shewale, M. A. D., & Sankpal, S. V. (2020). IOT based smart and secure health care system analysis & data comparison. *International Journal for Research in Applied Science and Engineering Technology*, *8*(1), 394–398.

[28] Mamoshina, P., Volosnikova, M., Ozerov, I. V., Putin, E., Skibina, E., Cortese, F., & Zhavoronkov, A. (2018). Machine learning on human muscle transcriptomic data for biomarker discovery and tissue-specific drug target identification. *Frontiers in Genetics*, *9*, 242.

[29] AlMotiri, S. H., Khan, M. A., & Alghamdi, M. A. (2016, August). Mobile health (m-health) system in the context of IoT. In *2016 IEEE 4th International Conference on Future Internet of Things and Cloud Workshops (FiCloudW)* (pp. 39–42). IEEE.

[30] Firouzi, F., Rahmani, A. M., Mankodiya, K., Badaroglu, M., Merrett, G. V., Wong, P., & Farahani, B. (2018). Internet-of-Things and big data for smarter healthcare: From device to architecture, applications and analytics. *Future Generation Computer Systems*, *78*, 583–586.

[31] Bharadwaj, H. K. et al. (2021). A review on the role of machine learning in enabling IoT based healthcare applications. *IEEE Access*, *9*, 38859–38890. doi:10.1109/ACCESS.2021.3059858.

[32] Pradhan, B., Bhattacharyya, S., & Pal, K. (2021). IoT-based applications in healthcare devices. *Journal of Healthcare Engineering,* vol. 2021, Article ID 6632599, 18 pages, 2021. https://doi.org/10.1155/2021/6632599

[33] Gupta, M., Thakur, N., Bansal, D., Chaudhary, G., Davaasambuu, B., & Hua, Q. (2022). CNN-LSTM hybrid real-time IoT-based cognitive approaches for ISLR with WebRTC: Auditory impaired assistive technology. *Journal of Healthcare Engineering, Hindawi*, 2022, ISSN: 2040–2309, Article ID 3978627, 17 pages, 21st Feb 2022. doi:10.1155/2022/3978627

[34] Kour, S., Kumar, R., & Gupta, M. (2021). Study on detection of breast cancer using Machine Learning. In *2021 International Conference in Advances in Power, Signal, and Information Technology (APSIT)*, 8–10 October 2021 (pp. 1–9).

Chapter 3

Telemedical and remote healthcare monitoring using IoT and machine learning

Sharad Chauhan, Kanika Pahwa, and Shakeel Ahmed

CONTENTS

3.1 Introduction 47
3.2 Literature review 50
3.3 Electronic healthcare monitoring system architecture 52
3.4 Healthcare monitoring system using IoT 53
3.5 Applications of IoT 58
3.6 Healthcare compatibility with machine learning 58
3.6.1 Supervised learning 58
3.6.2 Unsupervised learning 59
3.7 Secure remote healthcare system 60
3.8 IoT in telemedicine 61
3.8.1 Use cases of IoT in telemedicine 62
3.9 Challenges of IoT in healthcare 62
3.10 Conclusion and future scope 64
References 64

3.1 INTRODUCTION

Healthcare plays an important role in the life of every individual as it helps diagnose diseases at early stages so it can be cured with a medical treatment. Affected parts of body can be identified using various high-tech equipment, like CT and MRI. With the tremendous increase in the population of the world, the challenges in healthcare are also increasing. With the increase in population, there is also a need to increase the number of medical professionals and medical equipment. To maintain the quality of healthcare, there is a need to minimize the burden of the healthcare system. IoT (Internet of Things) is the ultimate solution for this problem. IoT devices allow real-time monitoring in case of medical emergencies, like heart attacks or asthma attacks [1]. Real-time monitoring means the data of a patient can be collected via a wearable device,

DOI: 10.1201/9781003305347-3

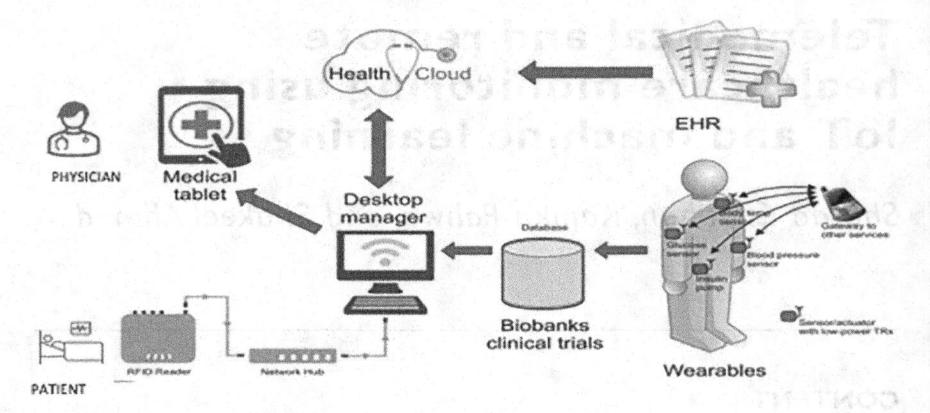

Figure 3.1 Role of IoT in hospitals [1].

and this data is shared using a data connection to a medical professional. Since a huge volume of data is being generated on a daily basis, data pre-processing, analysis and storage should be kept in mind, too. IoT helps reduce healthcare costs, as extra visits to the hospital would be decreased. Various clustering techniques using IoT are also helpful in healthcare systems [2].

Figure 3.1 shows the role of IoT in hospitals [1]. An asthma patient will retain an identification card that will contain all the required information related to their health (previous and newly scanned), including lab results, prescriptions and all other necessary information. All the details of the patient included in the ID card will be upload on the cloud so that doctors, nurses or physicians can access that information on their respective devices (laptops, computers, tablets or phones). This advancement in technology will be environment-friendly and can reduce or replace paperwork, and things will be totally digitized. These technological changes will act as a boon, as this global data will be securer, more efficient, more manageable and less error-prone and, thus, will save lives.

Machine learning [3] is a part of artificial intelligence which has a capability to improve medical process by training a model on the basis of past experience. Machine learning is used to implement different models to predict the disease prior to its occurrence. With the increase in data, it becomes difficult to analyse voluminous data manually. So machine learning algorithms are used for analysis. Machine learning algorithms are used to predict results from data which will help medical experts to make decisions. Researchers are now working to take machine learning to ICUs. The only objective is to extract a pattern from medical data to make healthcare better so that actionable decisions can be predicted.

It can be categorized into the following subcategories:

- **Healthcare**: Machine learning is widely used in the field of healthcare. With the increase in data, it becomes difficult to analyse voluminous data manually. So machine learning algorithms are used for analysis. Machine learning algorithms are used to predict results from data which will help medical experts to make decisions. Researchers are now working to take machine learning to ICUs. The only objective is to extract a pattern from medical data to make healthcare better so that actionable decisions can be predicted [4].

- **Marketing:** Nowadays, machine learning and marketing go hand in hand. Machine learning algorithms are used for marketing to make better profits. Using machine learning algorithms, we can target audience by analysing what type of keyword and phrases are most relevant.
- **Financial services:** Machine learning algorithms play a vital role to analyse and prevent unlawful activities. Along with this, machine learning is used by financial organizations to manage customer data. Numerous tools and techniques are available to transform data to information which is further used to make smart solutions.
- **Prediction while traveling:** While traveling from one point to another, machine learning can be used to predict the time taken by a vehicle on the basis of previous data.
- **Social media services:** Analysing data on social media is a challenging problem. Data on social media is highly unstructured, as there is no such rule and template for posting any information. To deal with this problem, machine learning algorithms are used to analyse voluminous data on social media. The better we can understand the audience of social media, the better we can make profits.
- **Extraction:** Machine learning algorithms are also used for extracting information from unstructured data. Information is available on web pages, social media, e-mails, business report or many more in different formats, which is extracted for making predictions or making smart decisions which will help organization in making profits.
- **Image and speech recognition:** Image and speech recognition is one of the most widely used real-time application of machine learning. In image recognition, each object is divided into pixels. Image recognition is used by Google in Google Photos, Google Drive and Google Search. While in speech recognition, an input voice note is converted to text. This application of machine learning is also used by Google in their Google Assistant, Google Translate and Google Text-to-Speech.
- **Suspicious activity detection from CCTVs:** Machine learning algorithms are also used to make society a safer place by detecting suspicious activities using CCTVs. The main idea is to analyse the data recorded by CCTV all the time, and in this way, it enables it to learn normal activities, like running, walking and talking. In case any notorious activity occurs in that area, let's say robbery, it provides an alert to the authority about the activity in real time.

3.2 LITERATURE REVIEW

Some related work has been done in the field of medicine with IoT [5] devices which are used to monitor a patient's health using wireless sensors. Some are outlined as follows:

- In [6], S. Laha et al. discussed the use of machine learning algorithm in IoT devices. IoT devices are embedded with sensors which are used for the collection of data which needs to be avoided from unauthorized access. In this paper, the authors discussed how machine learning can be used in wireless networks to control network traffic or for various prediction in the network.
- In [7], N. Taimoor and S. Rehman performed a survey of personalized healthcare services by analysing its key requirements, along with a three-layer architecture using IoT. They also proposed a reliable, resilient IoT-based personalized healthcare service. In this paper, the authors discussed the application of IoT in different fields that healthcare has used in maximum till today. In today's era, healthcare is developing towards personalization and digital wellness.
- In [8], S. Lavanya et al. focused on the design and implementation of an IoT-based health monitoring system which can be used for emergency situations. This approach reduces extra visits of the patient to the doctor, and the patient can directly be in touch with the doctor using IoT devices. This is used to check whether the patient is taking proper medication per a real-time clock (RTC). Data collection is done by sensors which are further connected to Raspberry Pi which is used to perform processing at some level.
- In [9], S. Thouheed Ahmed and M. Sandhya proposed a model which is further divided into two components, where one is used to provide remote monitoring in real time, and another is a diagnostic module which is used to send an early reminder for the case of two disorders, hypertension and hypotension. The diagnostic module uses the concept of fuzzy logic for early diagnosis of disease.
- In [10], Sarfraz Fayaz Khan discussed the different ways to improve the healthcare system for patients as well as doctors using sensors. The author discussed a four-layer healthcare monitoring system using IoT which includes a sensor layer, network layer, internet layer and service layer. In this paper, the author fused the hardware component of an IoT with the Wi-Fi module of RFID and other nodes. The output is clearly dependent on the RFID-based monitoring of data.
- In [11], C. Bhatt et al. proposed that IoT is used in the medical field to move healthcare systems to the next level with the advancement of technologies. IoT includes hardware components, like sensors,

embedded system and network connectivity, that interact with each other. This framework introduced by the author includes data collection using sensors, data processing and data storage. This process enables a patient to use the medical service anytime and anywhere per the requirement. IoT makes healthcare a smart healthcare system.

- In [12], P. Chatterjee et al. proposed the use of IoT in healthcare, which has brought a drastic change in healthcare. In this paper, the authors focused on IoT and decision support system to make the healthcare system a huge success by improving the quality of medical services and reducing the overall cost. Intelligent decision support system has been used to predict cardiovascular disease. IoT is used in healthcare to achieve goals which includes utilizing resources in the best way, making a better experience for patients and reducing the cost. A decision support system usually consists of three main components – inference engine, knowledge base and communication system.
- In [13], K. Chopra et al. discussed how IoT-enabled machine-to-machine communication can be further connected to the network for efficient communication. Different technology used with IoT are discussed in this paper, including RFID tag and different sensors used for data collection. In this, the authors discussed the use of IoT in different fields, like smart cities, smart home, prediction and smart transport system. Along with the benefits and its application, the authors discussed the challenges of implementing IoT which includes the requirement to standardize, confidentiality and network security.
- In [14], S. Syed Thouheed Ahmed et al. proposed a model to monitor people from rural areas who are mainly suffering from hypotension or hypertension. Data collected from mobile is analysed using biomedical sensors to take precautions in emergency situations. The authors also discussed the android application which will further capture data from the device itself and provide remote monitoring.
- In [15], K. Shailaja et al. discussed the different machine learning algorithm which is used in the field of healthcare to make efficient decisions. The authors discussed the different machine learning techniques which can be used to implement a decision support system, like SVM, decision tree, naïve Bayes, fuzzy logic and CART.
- In [16], D. Sherawat and N. S. Gill discussed the different types of sensors and how it can be used in the field of research. In this paper, the author discussed the position sensor, chemical sensor, proximity sensor, velocity sensor and their contribution to make the environment smart, like smart lighting, smart home, smart parking and many more applications.

3.3 ELECTRONIC HEALTHCARE MONITORING SYSTEM ARCHITECTURE

Here, in this section, we will be discussing about the architecture of electronic healthcare system. The architecture includes four layers, namely the data layer, domain layer, application layer and presentation layer. The role of each layer is discussed as follows:

- **Data layer**: The data layer of the proposed architecture shown in Figure 3.2 interacts with both the domain layer and application layer. This interaction will happen in such a way that only one layer can interact with the data layer at a time per the situation. If a patient or medical professional just needs to retrieve data for their information purpose, only then can the application layer directly communicate with the data layer, but if data is required for treatment purposes, then for that scenario, the application layer will interact with the domain layer first, and further, the domain layer will interact with the data layer. This process is followed especially when it is required to make appropriate decisions on the basis of inputs.
- **Domain layer**: This layer acts as a decision-making component of this architecture. This module consists of a logic which is used to make appropriate decisions. This is further divided into two modules, namely the inference engine and knowledge base. A knowledge base is a collection of rules and facts for different diseases. It is used to contain rules in the form of an if condition and a then action. The data which is taken from a user is taken as input data, and the action done on a given input will be considered as output or may be taken as input of another rule.
- **Application layer**: This layer is used to manage the data. All information, like the login credentials, regular information of the patient or assigned doctor, graph of the reports and interpretation diagrams, are managed by this layer. Along with this, it is also used to give reminders for regular check-ups and diagnosis to the patient or to doctors.
- **Presentation layer**: This layer is used as an interface for the doctor or patient. As users can access via a mobile or a system, the interface should be designed in such a way that both system and mobile users should be able to access it efficiently. This interface will be complicated on the patient's side, as it needs to be attached to some medical device, and as a result, they need some graphs for quick and easy interpretations.

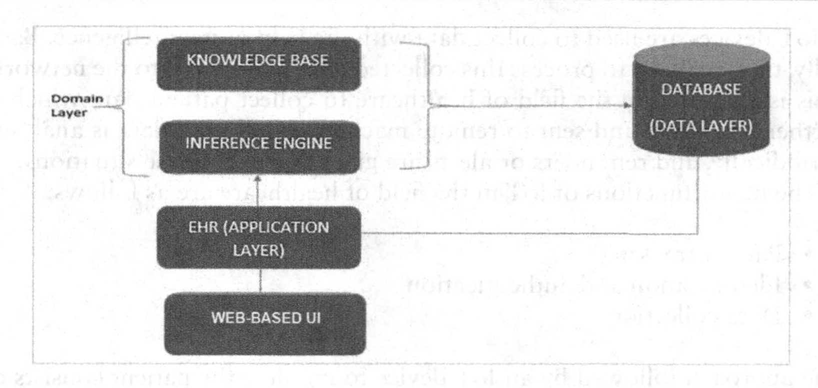

Figure 3.2 Architecture of an EHMS [18].

3.4 HEALTHCARE MONITORING SYSTEM USING IOT

The Internet of Things is a network of devices with some intelligence. Objects involved in IoT may have the capability to interact, or it may be used to just collect data [17]. An IoT network is used to hold a huge amount of data. The IoT can be represented as different layers where each layer has its own functionalities:

- **Acceptance layer:** This layer includes sensing devices which are used to collect data and pass the collected data to an upper layer for further analysis. The different sensors used are Zigbee, RFID and all kinds of sensors.
- **Network layer:** The main responsibility of this layer is to provide efficient communication and secure data transfer. Advanced technologies have been used for wireless communication.
- **Intermediate layer:** This layer is used to store data safely in some predefined format. This can be further used by other layers for service management.
- **Application layer:** The main responsibility of this layer is the provision of all types of services specific to the field.
- **Business layer:** This layer is mainly used for analysis purposes, and it helps in making appropriate decisions.

Traditional healthcare systems used to have a continuous diagnosis of a patient which may result in a wrong diagnosis, inappropriate decision-making and the loss of data related to patient history, so IoT solves all these problems by introducing IoT-based healthcare.

IoT devices are used to collect data with the help of its intelligence. Basically, they are used to process this collected data and send it to the network. This is also used in the field of healthcare to collect patient data which is further processed and sent to remote machines. Collected data is analysed periodically, and reminders or alerts are given for worst-case situations.

The major functions of IoT in the field of healthcare are as follows:

- Patient tracking
- Identification and authentication
- Data collection

The approach followed by an IoT device to monitor the patient consists of four stages, as shown in Figure 3.3:

- **Data collection using sensors:** Different sensors are used by the user to collect data, like heartbeat rate, BP level, blood glucose level or temperature, as shown in Figure 3.4.
- **Data processing:** It includes the data display, temporary data storage or initial data processing by Raspberry Pi [18].
- **Data storage:** After the further processing of data, it is forwarded to the IoT server for storage.
- **Data transmission:** Data stored on the IoT server is accessed by other clients involved in the system to give medical advice to the required user.

The data of the patient is collected using sensors on the Raspberry Pi. This collected data is stored on an IoT server with the patient's unique ID and

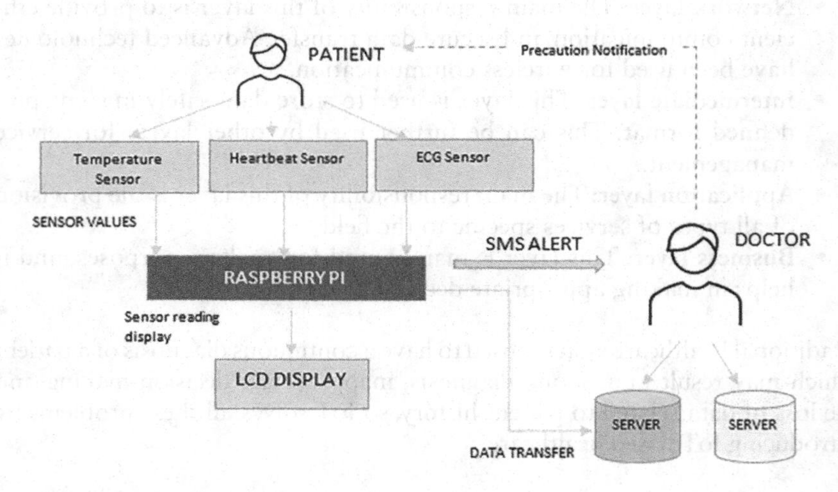

Figure 3.3 The approach followed by an IoT device [19].

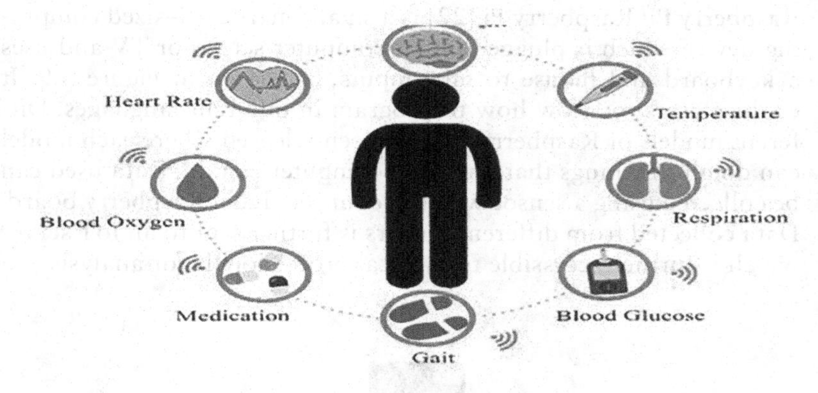

Figure 3.4 Wearable IoT devices.

passed to medical professionals for analysis. The sensing device chosen by the patient depends on the parameter which the patient wants to be monitored by doctors. But parameter like body temperature or heartbeat rate are the parameters which are monitored continuously.

Different devices used to implement IoT in healthcare are as follows:

- **Temperature sensor:** This sensor [20] is used to measure temperature. The LM35 temperature sensor shown in Figure 3.5 is said to be most accurate per the requirement of the healthcare field. In this sensor, output voltage varies per the variation in temperature [21]. For every 1 degree Celsius rise, there is a 0.01V increase in the voltage.
- **ECG sensor:** It is used to monitor heartbeat rate and other disease indications. A basic threshold value is 200ms; if the value reaches below the threshold, then an alarm is generated. The ECG sensor is shown in Figure 3.6.
- **Pulse rate sensor:** A pulse sensor is mainly used to compute the pulse rate. As shown in Figure 3.7, the pulse rate sensor has a heart-shaped logo on the front. This side is used to be in contact with the skin to calculate the pulse rate. On the front side, there is one LED and light sensor which is used to flash lights per the brightness. The rest of the parts is mounted on the back part of the sensor.
- **Blood pressure sensor:** A blood pressure sensor is used to measure blood pressure, and it is automatic and very easy to operate. It is very compact and easy to fit on the wrist, similar to a watch, as shown in Figure 3.8. Usually, medical professionals used to manually press the cuff which is further linked to a mercury column to increase the blood pressure. But this pressure sensor enables the calculation of blood pressure without the need of mercury.

- **Raspberry Pi:** Raspberry Pi [22] is a small smart card–sized computing device which is plugged into a computer screen or TV and uses a keyboard and mouse to take inputs, as shown in Figure 3.9. It enables users to know how to program in different languages. Different models of Raspberry Pi have been released where each model can do all the things that a normal computer can do. Data used can be collected using a sensor which is connected to a Raspberry board. Data collected from different sensors is further sent to an IoT server which is further accessible to medical professionals for analysis.

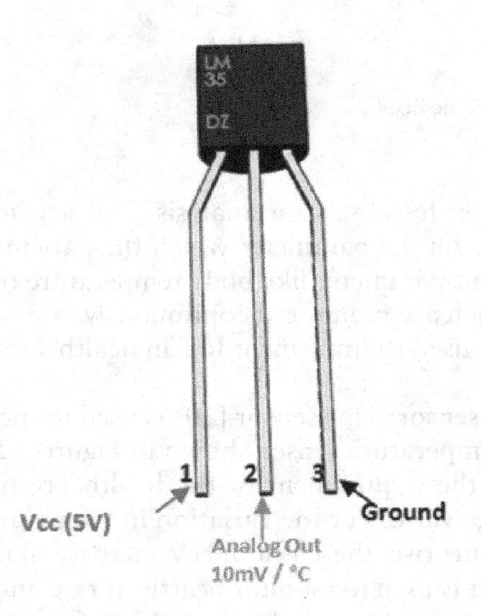

Figure 3.5 Temperature sensor [20].

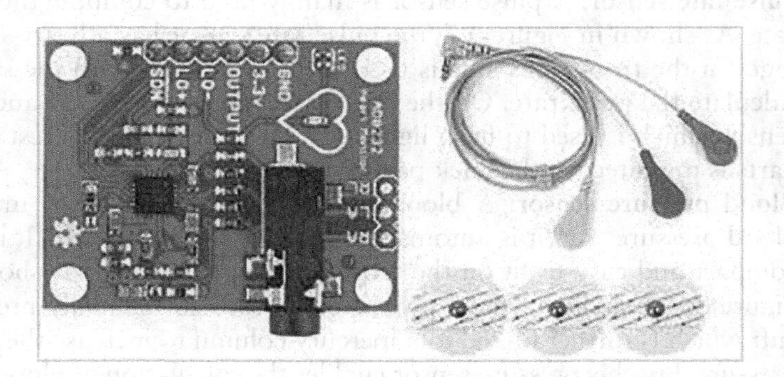

Figure 3.6 ECG sensor.

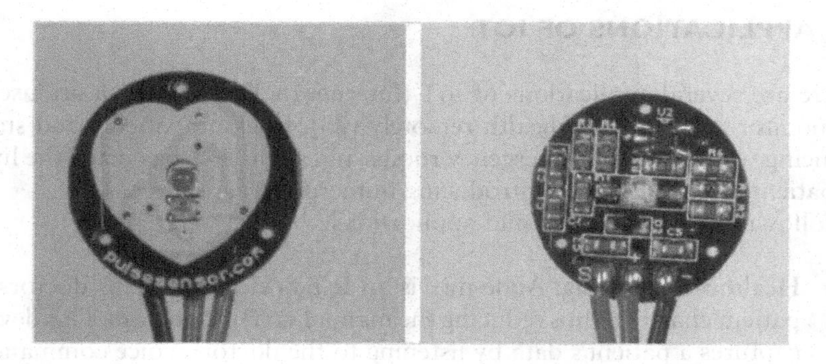

Figure 3.7 Pulse rate sensor [20].

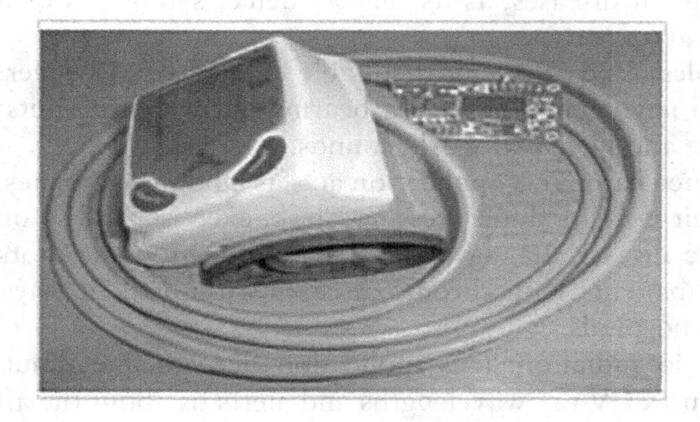

Figure 3.8 Blood pressure sensor.

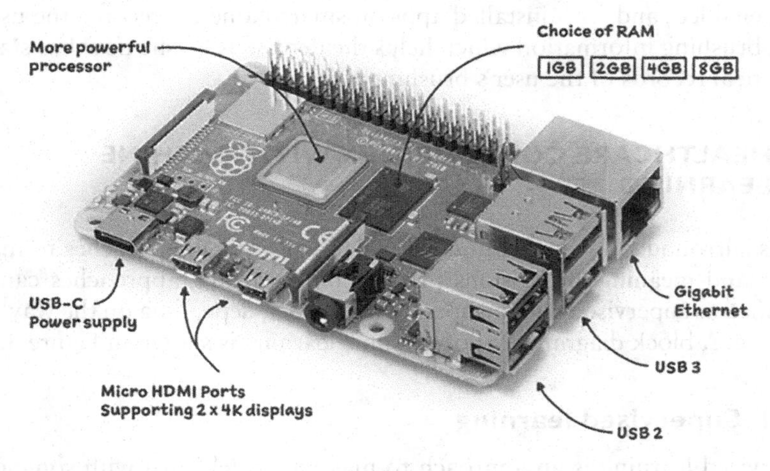

Figure 3.9 Raspberry Pi [22].

3.5 APPLICATIONS OF IOT

There are several applications of IoT (Internet of Things) which are useful in monitoring a patient's health remotely [23], tracking patients and staff, reducing waiting time in emergency rooms and so on. IoT has made the lives of patients comfortable by introducing numerous wearables [24].

Following are the devices and applications:

- **Healthcare charting:** Aude-mix is an IoT device that helps doctors in patient charting, thus reducing the manual work of doctors. This device captures a patient's data by listening to the doctor's voice commands.
- **Ingestible sensors:** Proteus Digital Health is an IoT-enabled device whose sensors can help diabetic patients by providing them with early warnings of diseases, as its sensors detect symptoms of the patient's disease at an early stage.
- **Hearables:** One of the most suitable examples is Doppler Labs's IoT-enabled new technology for a hearing aid which permits the user to equalize and filter real-world sounds.
- **Sportsmen care:** This application acts as a boon to athletes, as it monitors their weight, blood pressure, sleep, exercise, and so on.
- **Chronic disease management:** This application monitors and takes care of patients with chronic diseases without their physical attendance in hospitals.
- **Ultraviolet radiation:** This application alerts people about UV rays as it measures UV ray wavelengths and alerts us about the affected areas or the time when they are more harmful.
- **Dental health:** This is one of the most interesting applications and can be used by everyone. It contains a toothbrush that is Bluetooth-enabled, and with installed apps on smartphones, it records the user's brushing information which helps the dentist as he/she now has statistical records of the user's brushing habits.

3.6 HEALTHCARE COMPATIBILITY WITH MACHINE LEARNING

IoT is surrounded by applications of machine learning approaches to make better and meaningful decisions [4]. Machine learning approaches can be classified as supervised and unsupervised learning, depending on the way it is learning. A block diagram [23] of machine learning is shown in Figure 3.10.

3.6.1 Supervised learning

Supervised learning is an approach to make a model learn with some corrected or labelled data. This data is used to train a model, and then the trained model is used for testing by feeding new data to it.

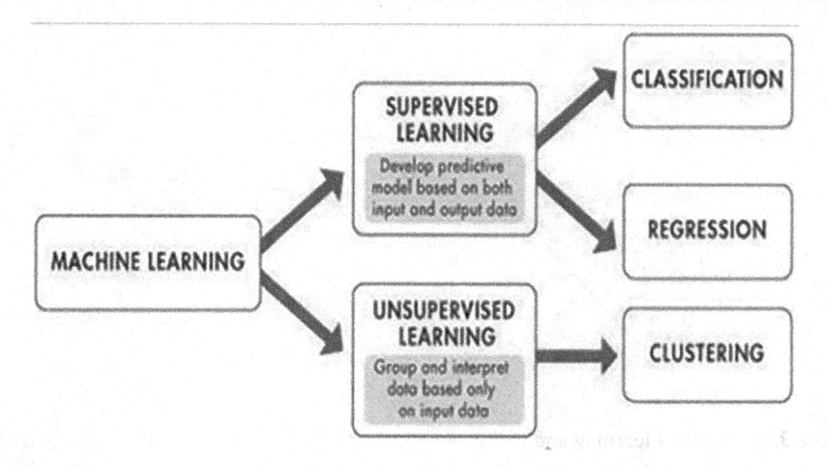

Figure 3.10 Machine learning approaches.

Supervised approaches are broadly classified into two categories:

- **Classification:** These models are used to classify the data into categories like pass or fail, true or false. Classification algorithms are decision tree algorithm, support vector machine, neural network or KNN.
- **Regression:** This approach is used in cases when output is assumed to be a real value. Algorithms under the category of regression are logistics regression, linear regression and multivariate regression.

3.6.2 Unsupervised learning

In the unsupervised machine learning approach, the machine learns on its own. No prior information is available like in the supervised approach. The prime objective of this approach is to divide the data into groups on the basis of its similarities.

Machine learning approaches are applied on the data collected by sensors to make decisions whether the data needs to be stored on the server or needs to be shared with medical professional urgently. After this, further ML approaches are applied on stored data to analyse the problem on the basis of the collected real data, and the response is shared with the patient by giving a prescription.

The healthcare system uses machine learning approaches to find patterns and extract abnormalities in data. Also, machine learning approaches are used for data visualization which further gives better understanding. These visualization reports are stored on the IoT server. Further, this data is used by all the users involved in the real-time monitoring of the patient. A broader view of using machine learning and IoT is represented in Figure 3.11.

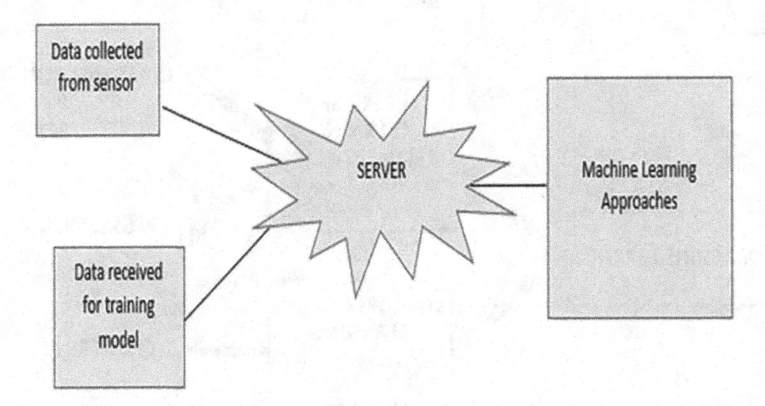

Figure 3.11 Machine learning and IoT.

3.7 SECURE REMOTE HEALTHCARE SYSTEM

As in the case of e-healthcare, there is a need to send the data collected from the IoT device to the cloud, so there may be a chance of the data being accessed by an unauthorized person which may affect its integrity and confidentiality. So a security mechanism needs to be implemented while transferring data [25]. For this purpose, several cryptographic algorithms are used which are broadly classified into symmetric cryptography, asymmetric cryptography and hashing. In these algorithms, digital keys are used to encrypt and decrypt the data. In the symmetric key algorithm, the same key is used for encryption and decryption; while in the asymmetric key cryptography, two different keys are used for encryption and decryption. The hashing algorithm is the most effective algorithm. Hash here means a unique number is generated from a text.

Encryption algorithms which can be used to encrypt the data are as follows (Table 3.1):

- **Data Encryption Standard (DES):** DES uses the same key for encryption and decryption. It uses a block of 64 bits and a key of 56 bits for the encryption process. This process of encryption is known as symmetric key encryption. Basically, it was developed by the US government to make all government communications secure. The main objective of this algorithm is to achieve security and standardization.
- **Advanced Encryption Standard (AES):** This algorithm also uses the same key for encryption but of varying length. It uses the process of re-encrypting the same block multiple times, depending on the value of the key length. The block size in this algorithm is 128, 192 or 256 bits, while the key length is of size 128, 192 or 256 bits. This algorithm is securer and faster.

Table 3.1 Comparison between different encryption algorithms [1]

Algorithm	*Block size*	*Key length*	*Security*	*Speed*
RSA	Variable length	1024–4096 bits	Considered secure	Very slow
DES	64 bits	56 bits	Inadequate	Very slow
3DES	64 bits	112 bits, 168 bits	Secure	Slow
AES [1]	128 bits	128,192, 256 bits	Highly secure	Very fast
Twofish	128 bits	[26] 6 bits	Secure	Very fast

- **Triple DES:** Triple DES is a symmetric key algorithm. Triple DES is an advancement of the DES approach, as in this algorithm, each block undergoes three phases of encryption. Three different keys are used for the encryption process. This algorithm is used by IoT products because of its extreme flexibility and compatibility. The block size used is of 64 bits, and the key length is 112 and 168 bits.
- **RSA encryption (Rivest, Shamir and Adleman):** RSA uses different sets of keys for encryption and decryption, and is known as an asymmetric key algorithm. This is applied on a block of size 128 bits with the key length of size 1024 bits to 4096 bits. A public key is used to encrypt the data. This encrypted data will be decrypted by a private key only.
- **Twofish encryption algorithm:** The Twofish algorithm is a symmetric key algorithm which is applied on a block size of 128 bits. The length of the key will be up to 256 bits. This approach has some distinctive features compared to other encryption algorithms. This approach has been used in computers with a lower capacity of processor and used in different encryption software as well.

3.8 IOT IN TELEMEDICINE

Remote monitoring of patients [26] helps patients and doctors be connected consistently even at a geographical distance. This process can be implemented using IoT [23]. IoT in telemedicine helps patients suffering from chronic diseases or old age. It is beneficial for situations when people are busy with their schedules and are struggling to visit physicians personally.

The role of IoT in telemedicine is shown in Figure 3.12. Here, a patient interacts with the IoT devices [27], which collect medical data in terms of heartbeat rate, body temperature and blood pressure. This data is passed to an IoT server after applying different security algorithms [31]. Further, doctors will monitor the parameters collected from sensors and make decisions based on the collected data. After diagnosing the problem, a medical professional can now write the prescription and forward this prescription to the medical pharmacy per the requirement. Further, the pharmacy can contact the patient for pickup.

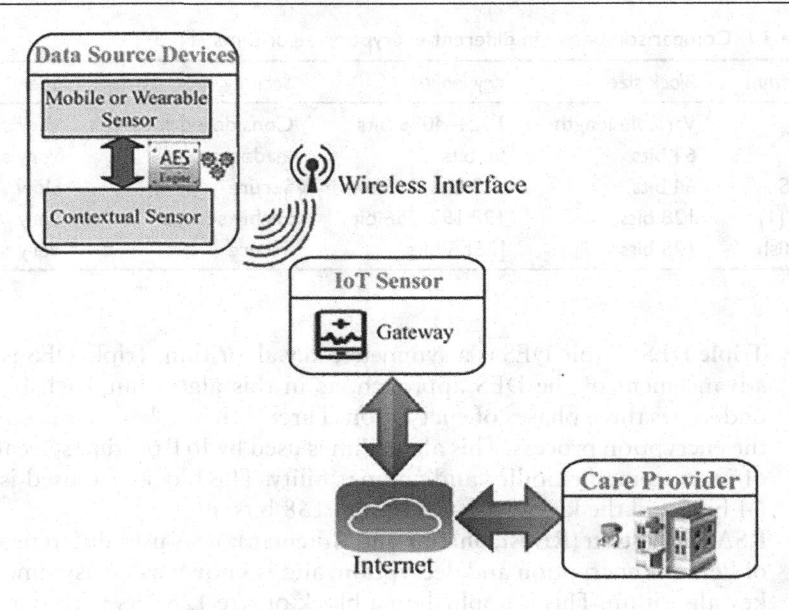

Figure 3.12 Role of IoT in telemedicine [23].

3.8.1 Use cases of IoT in telemedicine

- **Cyber Med:** This is basically a blend of IoT and telemedicine. Patient used to buy a digital stethoscope or an oximeter for a virtual appointment. The data collected using sensors is transmitted to the cloud so that it can be accessed by the physician for evaluation.
- **Pillsy:** It is a smart medication bottle which is used to give reminders to the patients to take medicine and to avoid overdose. It is used to keep a record of the medication period and refill it by ordering it.
- **Mobile medicine:** It is used to measure parameters like heartbeat rate, body temperature and blood sugar level for each registered patient. A medical profile is maintained of each patient with specific IDs. Based on the collected data, it is used to analyse and, as a result, gives an indicator to the patient or their relative. This will make the patient focus on their diets and regular exercises. It also generates advice from medical professionals on a regular basis.

3.9 CHALLENGES OF IOT IN HEALTHCARE

IoT has brought a revolutionary change in the field of healthcare. Along with all the benefits, there are some challenges in implementing IoT in

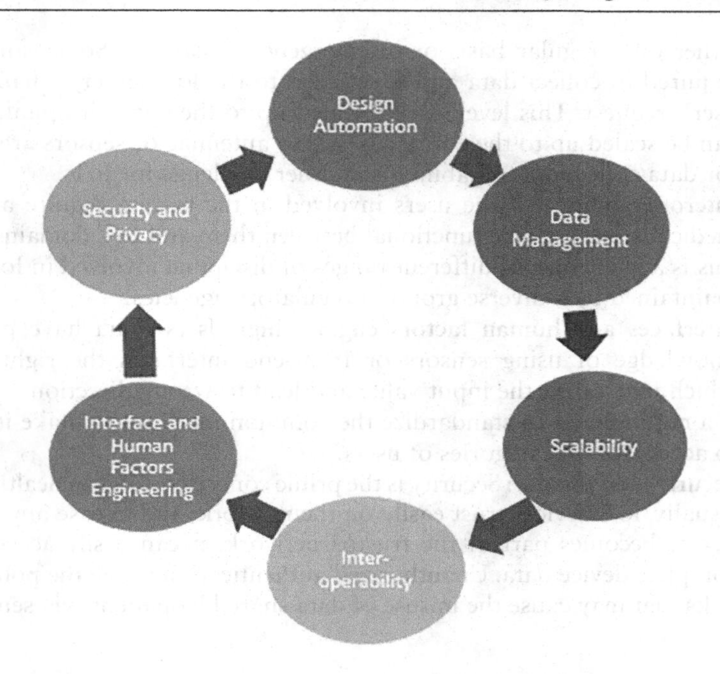

Figure 3.13 Challenges of IoT [30].

healthcare, as shown in Figure 3.13. Some of the challenges of IoT are discussed subsequently:

- **Design automation:** IoT in healthcare involves both the cyber and physical world. It is the combination of hardware and software components, along with the network. So, it is required to focus on the compatibility of a heterogeneous environment [28].
- **Data management:** Data management and storage are some of the biggest challenges of IoT in healthcare. As data is being collected from wearable devices which is affected with time. So it becomes a great challenge to deal with this voluminous data. Another issue related to data, along with data volume, is data variety. As different IoT devices are used to collect data in different forms, data management will become difficult in that scenario, like in case of ECG [29] data collected in the form of XML, while in the case of camera-based IoT devices, data will be in an image format.
- **Scalability:** Scalability is one of the key requirements of IoT. If e-healthcare is developed in smaller scale, then every registered patient can easily have direct access to all the medical services which are required

either on a regular basis or in emergency situations. So sensors are required to collect data and forward it to the IoT server to handle a user's request. This level can be scaled up to the entire hospital, or it can be scaled up to the entire city where antennae or sensors are used for data collection. Scalability is another challenge for IoT.

- **Interoperability:** All the users involved in the system require all the medical services to be functional between them and IoT domains. But this is a challenge, as different ranges of discipline involved in IoT are maintained by a diverse group of regulatory agencies.
- **Interfaces and human factors engineering:** Users don't have proper knowledge of using sensors or front-end interfaces the right way which may affect the input value and lead to wrong directions. There is a requirement to standardize the common interface to make it easy to access for all categories of users.
- **Security and privacy:** Security is the prime concern for IoT in healthcare. Usually, IoT devices trust easily on the network, and in case any other device becomes part of the trusted network, it can easily access the complete device data. Unauthorized authentication is one the potential risks that may cause the misuse of data shared by patients via sensors.

3.10 CONCLUSION AND FUTURE SCOPE

This chapter discussed the use of IoT and machine learning in healthcare. IoT is the solution to many challenges in the medical field, as it enables real-time monitoring of patients suffering from chronic diseases. Machine learning approaches are used to make decisions while analysing the data. Remote healthcare will reduce the rounds of visits of the patients in hospitals, as they will be able to give their medical information via sensors, and this data is retrieved by doctors for their medical services. In this chapter, the architecture of IoT was discussed, along with the different sensors used to collect data. Use cases of e-healthcare were also included in this chapter.

REFERENCES

[1] N. Agnihotri, and A. K. Sharma, "Comparative analysis of different symmetric encryption techniques based on computation time," *PDGC 2020 – 2020 6th Int. Conf. Parallel, Distrib. Grid Comput.*, pp. 6–9, 2020, doi:10.1109/PDGC50313.2020.9315848.

[2] Sharad, S. S. Kang, and Deepshikha, "Cluster based techniques leach and modified LEACH using optimized technique EHO in WSN," *Int. J. Innov. Technol. Explor. Eng.*, vol. 8, no. 9 Special Issue, pp. 363–372, 2019, doi:10.35940/ijitee.I1058.0789S19.

[3] S. Ray, "A Quick Review of Machine Learning Algorithms," *Proc. Int. Conf. Mach. Learn. Big Data, Cloud Parallel Comput. Trends, Prespectives Prospect. Com. 2019*, pp. 35–39, 2019, doi:10.1109/COMITCon.2019.8862451.
[4] S. Chauhan, R. Arora, and N. Arora, "Researcher issues and future directions in healthcare using IoT and machine learning," in *Smart Healthcare Monitoring Using IoT with 5G*, Ist., G. C. Meenu Gupta and V. H. C. de Albuquerque, Eds. CRC Press, Taylor and Francis Group, Boca Raton, Londan, New York, 2021, pp. 177–196.
[5] A. Hosseinian-Far, M. Ramachandran, and C. L. Slack, "Emerging trends in cloud computing, big data, fog computing, IoT and smart living," *Technol. Smart Futur.*, pp. 29–40, 2017, doi:10.1007/978-3-319-60137-3_2.
[6] S. Laha, N. Chowdhury, and R. Karmakar, "How can machine learning impact on wireless network and IoT? – A survey," *2020 11th Int. Conf. Comput. Commun. Netw. Technol. ICCCNT 2020*, 2020, doi:10.1109/ICCCNT49239.2020.9225652.
[7] N. Taimoor, and S. Rehman, "Reliable and resilient AI and IoT-based personalised healthcare services: A survey," *IEEE Access*, vol. 10, pp. 535–563, 2022, doi:10.1109/ACCESS.2021.3137364.
[8] S. Lavanya, G. Lavanya, and J. Divyabharathi, "Remote prescription and I-Home healthcare based on IoT," *IEEE Int. Conf. Innov. Green Energy Healthc. Technol. – 2017, IGEHT 2017*, pp. 1–3, 2017, doi:10.1109/IGEHT.2017.8094069.
[9] S. Thouheed Ahmed, and M. Sandhya, *Real-Time Biomedical Recursive Images Detection Algorithm for Indian Telemedicine Environment*, vol. 768, no. October 2018. Springer, Singapore, 2019.
[10] S. F. Khan, "Health care monitoring system in Internet of Things (IoT) by using RFID," *2017 6th Int. Conf. Ind. Technol. Manag. ICITM 2017*, pp. 198–204, 2017, doi:10.1109/ICITM.2017.7917920.
[11] C. Bhatt, N. Dey, and A. S. Ashour (eds.) *Internet of Things and Big Data Technologies for Next Generation Healthcare*, vol. 23, 2017, ISBN: 978-3-319-49735-8.
[12] P. Chatterjee, L. J. Cymberknop, and R. L. Armentano, "IoT-based decision support system for intelligent healthcare: Applied to cardiovascular diseases," *Proc. - 7th Int. Conf. Commun. Syst. Netw. Technol. CSNT 2017*, pp. 362–366, 2018, doi:10.1109/CSNT.2017.8418567.
[13] K. Chopra, K. Gupta, and A. Lambora, "Future internet: The internet of things—A literature review," *Proc. Int. Conf. Mach. Learn. Big Data, Cloud Parallel Comput. Trends, Prespectives Prospect. Com. 2019*, pp. 135–139, 2019, doi:10.1109/COMITCon.2019.8862269.
[14] S. Syed Thouheed Ahmed, K. Thanuja, N. S. Guptha, and S. Narasimha, "Telemedicine approach for remote patient monitoring system using smart phones with an economical hardware kit," *2016 Int. Conf. Comput. Technol. Intell. Data Eng. ICCTIDE 2016*, pp. 7–10, 2016, doi:10.1109/ICCTIDE.2016.7725324.
[15] K. Shailaja, B. Seetharamulu, and M. A. Jabbar, "Machine learning in healthcare: A review," In *2018 Second Int. Conf. Electron. Commun. Aerospace Technol. (ICECA)*, IEEE, 2018 March, pp. 910–914.
[16] D. Sehrawat, and N. S. Gill, "Smart sensors: Analysis of different types of IoT sensors," *Proc. Int. Conf. Trends Electron. Informatics, ICOEI 2019*, no. Icoei, pp. 523–528, 2019, doi:10.1109/ICOEI.2019.8862778.

[17] J. C. Kim, and K. Chung, "Hybrid multi-modal deep learning using collaborative concat layer in health bigdata," *IEEE Access*, vol. 8, pp. 192469–192480, 2020, doi:10.1109/ACCESS.2020.3031762.
[18] B. I. E. Ahmad, M. A. Ayu, I. Abdullahi, and Y. Yakubu, "Remote patient monitoring system architecture for diabetes management," *3rd Int. Conf. Comput. Eng. Des. ICCED 2017*, vol. 2018-March, pp. 1–6, 2018, doi:10.1109/CED.2017.8308120.
[19] A. Gutte, and R. Vadali, "IoT based health monitoring system using Raspberry Pi," *2018 Fourth International Conference on Computing Communication Control and Automation (ICCUBEA)* (pp. 1–5). IEEE, 2018 August.
[20] Sharad, E. N. Kaur, and I. K. Aulakh, "Evaluation and implementation of cluster head selection in WSN using Contiki/Cooja simulator," *J. Stat. Manag. Syst.*, vol. 23, no. 2, pp. 407–418, 2020, doi:10.1080/09720510.2020.1736324.
[21] H. N. Saha, D. Paul, S. Chaudhury, S. Haldar, and R. Mukherjee, "Internet of Thing based healthcare monitoring system," *2017 8th IEEE Annu. Inf. Technol. Electron. Mob. Commun. Conf. IEMCON 2017*, pp. 531–535, 2017, doi:10.1109/IEMCON.2017.8117245.
[22] S. Tanwar, P. Patel, K. Patel, S. Tyagi, N. Kumar, and M. S. Obaidat, "An advanced internet of thing based security alert system for smart home," *IEEE CITS 2017 – 2017 Int. Conf. Comput. Inf. Telecommun. Syst.*, pp. 25–29, 2017, doi:10.1109/CITS.2017.8035326.
[23] U. Albalawi, and S. Joshi, "Secure and trusted telemedicine in Internet of Things IoT," *IEEE World Forum Internet Things, WF-IoT 2018 - Proc.*, vol. 2018-January, pp. 30–34, 2018, doi:10.1109/WF-IoT.2018.8355206.
[24] T. Saimounika, and M. Tech, "Internet of things," *2017 Int. Conf. Energy, Commun. Data Anal. Soft Comput.*, no. December 2019, pp. 3507–3508, 2017.
[25] J. Vora *et al.*, "Ensuring privacy and security in E-health records," *CITS 2018 - 2018 Int. Conf. Comput. Inf. Telecommun. Syst.*, 2018, doi:10.1109/CITS.2018.8440164.
[26] V. Tamilselvi, S. Sribalaji, P. Vigneshwaran, P. Vinu, and J. Geetha Ramani, "IoT based health monitoring system," *2020 6th Int. Conf. Adv. Comput. Commun. Syst. (ICACCS)*, IEEE, 2020. pp. 386–389.
[27] I. A. Pap, S. Oniga, I. Orha, and A. Alexan, "IoT-based eHealth data acquisition system," *2018 IEEE Int. Conf. Autom. Qual. Testing, Robot. AQTR 2018 - THETA 21st Ed. Proc.*, pp. 1–5, 2018, doi:10.1109/AQTR.2018.8402711.
[28] F. Ahamed, and F. Farid, "Applying internet of things and machine-learning for personalized healthcare: Issues and challenges," *Proc. - Int. Conf. Mach. Learn. Data Eng. iCMLDE 2018*, pp. 22–29, 2019, doi:10.1109/iCMLDE.2018.00014.
[29] A. D. Acharya, and S. N. Patil, "IoT based health care monitoring kit," *Proc. 4th Int. Conf. Comput. Methodol. Commun. ICCMC 2020*, no. Iccmc, pp. 363–368, 2020, doi:10.1109/ICCMC48092.2020.ICCMC-00068.
[30] F. Firouzi, B. Farahani, M. Ibrahim, and K. Chakrabarty, "Keynote paper: From EDA to IoT eHealth: Promises, challenges, and solutions," *IEEE Trans. Comput. Des. Integr. Circuits Syst.*, vol. 37, no. 12, pp. 2965–2978, 2018, doi:10.1109/TCAD.2018.2801227.
[31] L. Fan, "Usage of narrowband internet of things in smart medicine and construction of robotic rehabilitation system," *IEEE Access*, vol. 10, pp. 6246–6259, 2021.

Chapter 4

Efficient ways for healthcare data management using data science and machine learning

Dimple Chehal and Payal Gulati

CONTENTS

4.1 Introduction 67
4.2 Big data definition 70
4.3 The healthcare domain's big data analysis process 72
 4.3.1 Data acquisition 73
 4.3.2 Data storage 73
 4.3.3 Data management 74
 4.3.4 Data analysis 74
 4.3.5 Data visualization 74
4.4 Big data applications in the healthcare domain 75
 4.4.1 Omics 75
 4.4.2 Insurance industry 75
 4.4.3 Pharmaceuticals 75
 4.4.4 Personalized patient care 75
 4.4.5 Hospital management 76
4.5 Challenges of big data analysis in healthcare 76
4.6 Conclusion 76
References 77

4.1 INTRODUCTION

The digital era is trying to cope up with the generation of a humongous amount of data. In 2017, the International Data Corporation (IDC) approximated the dimension of the digital universe as 16 zettabytes (ZB) and 130 exabytes (EB) in 2020. This illustrates the speed at which the universe is expanding digitally. Depending on user preferences, a variety of user-related information, such as location, applications downloaded, calorie intake, internet browsing history, contacts, e-mails, calls and purchase patterns, are stored by internet giants. Analysing critical data can lead to a significant increase in

DOI: 10.1201/9781003305347-4

the generation of revenues and improve the user experience [1]. The advancements in technology have generated a huge amount of data to an extent that management of such data is not possible with traditional techniques. In order to meet the growing demand in healthcare industries, new strategies are required to organize, manage and analyse this big data. Big data (BD) is characterized according to three Vs, namely, volume, velocity and variety [2,3]. Volume denotes the existence of massive data, whereas velocity denotes the pace at which it is created, and diversity denotes a wide range of data types. Veracity means the quality of data, and the value of the data has been accepted as the fourth and fifth definition of big data, respectively [4,5].

Healthcare management aims to diagnose, treat and prevent health-related issues in people. The stakeholders involved include the patients, professionals (such as doctors and nurses), pharma and clinical researchers, health facilities (such as clinics, public healthcare (PHC) centres and hospitals) and financing institutions which intermediate the financial transactions between them [6,7]. Previously, all the information related to a patient's healthcare, such as medical history, laboratory test data and personal information, was stored in an offline way, in typed or handwritten records. With the advent of the digital era, the user can access all his/her healthcare data at the click of a button, as the data is available electronically [8,9]. The role of big data in healthcare is to identify new data sources along with legacy or traditional data sources so as to attain novel solutions to various health-related problems. The insights gained from these sources can transform the way personalized and cost-effective services are offered to the patients. This is the reason healthcare management has become patient-centred from disease-centred. The Google trend of web search of big data in healthcare over the last decade is shown in Figure 4.1. Figure 4.2 shows the top 5 regions searching big data in healthcare, with Singapore and India holding the first and second position, respectively.

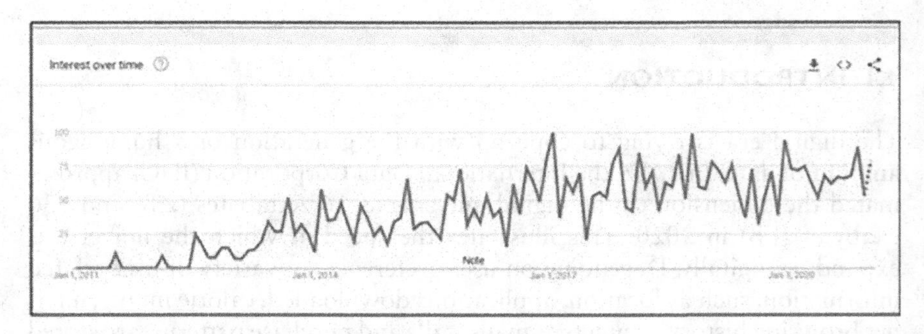

Figure 4.1 **Trend of big data web search for healthcare by Google.**

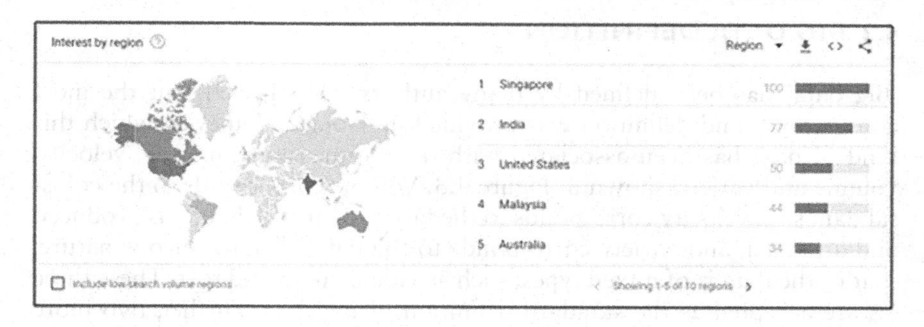

Figure 4.2 Region-wise search interest on big data for healthcare.

Hospital records, patients' medical history, medical examination results and the various machines and devices used to examine patients are part of the internet of things and are the sources of big data in the healthcare industry. These sources can be broadly categorized as follows [5]:

i. *Structured data:* Such data adheres to a strict format, data type or a structure. For instance, terminologies of diseases, its symptoms, medical test results and patient billing information form structured data.
ii. *Semi-structured data:* Data that follows just a minimum format and is self-explanatory in nature. For example, data captured from sensors of undermonitored patients.
iii. *Unstructured data:* Data that doesn't follow any format, such as medical prescription by doctors in a natural language.

While big data in healthcare can revolutionize the way diseases are predicted [11], help in analysis of the medicine intake by a patient in his lifetime and provide personalized solutions to the patients' ailments, there exists several challenges which need to be addressed for the revolution to actually be in place [12]. The challenge is to equip healthcare providers with an infrastructure to capture and analyse big data. Additionally, to formulate an accurate and useful model, the model should be well calibrated and updated from time to time by clinical experts. Not only this, the prediction model should be capable of explaining its outcome and have broad applications in real-life scenarios [7,13,15].

The rest of the chapter is detailed as follows: Section 4.2 briefs the dimensions and terminologies used in big data with respect to healthcare domain. Section 4.3 focuses on the processes followed by the authors to study the available literature on healthcare big data. Further, Section 4.4 discusses big data and the associated healthcare processes. The application, advantages and disadvantages are detailed in Section 4.5. Lastly, Section 4.6 summarizes the investigation by suggesting further recommendations in this area.

4.2 BIG DATA DEFINITION

'Big data' has been defined by many authors [9,16,18,19], but the most commonly found definition was provided by Douglas Laney in which this kind of data has been associated with three dimensions, namely, velocity, volume and variety, shown in Figure 4.3. Volume corresponds to the colossal data size, velocity corresponds to the fast pace at which data is produced and analysed, and variety corresponds to the data's heterogeneous nature, that is, the data is of mixed types, such as video, audio and text. These three Vs are accepted as the standard definition of big data. Further, two more Vs have been added to this definition, namely veracity and value. While the former indicates data reliability and quality, the latter indicates gaining of hidden insights from data analysis [20–22]. Big data in healthcare requires collecting, storing, processing, managing, analysing and visualizing this data.

The amount of global healthcare data has increased to 2314 exabytes in 2020. The amount of global healthcare [23] data generated in 2013 and a projection for 2020 is shown in the Figure 4.5. Thus, day by day, the

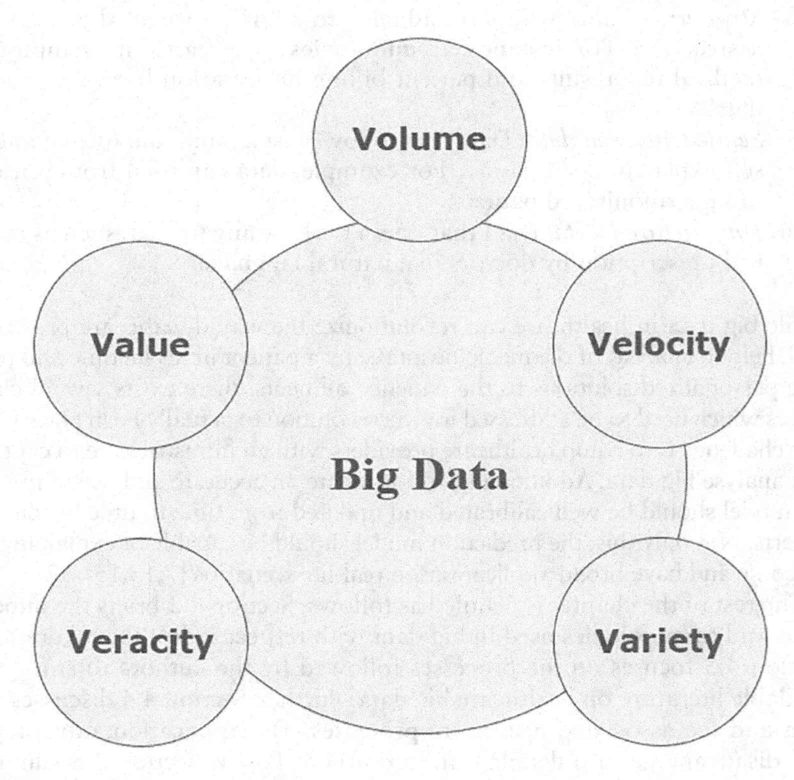

Figure 4.3 Characteristics of big data.

increasing healthcare data is creating issues for storage and analysis. The colossal size of data implies a large amount of data in terabytes (10^{12} bytes) or larger, as shown in Figure 4.4. Data in the form of clinical records, experiments results, patient metadata and healthcare staff data might grow up to zettabytes (10^{21} bytes) or yottabytes (10^{24} bytes) in the near future. Such kinds of data are beyond the handling capacity of traditional data management systems which led to the idea of the development of big data management systems.

The data in context can be organized or structured according to predefined formats, as in the case of patient(s)' reports, or can be unorganized or unstructured, just like a patient's handwritten prescription by a doctor. Semi-organized is a mix of the two types previously mentioned.

Velocity corresponds to the massive speed of data generation and its management. For instance, velocity is the speed at which a new viral video is circulated among people in their online social circles. Variety of data is due to the fact that data in question is not of the same type, and instead, this data arises from different types of sources. The completeness and accuracy of information are referred to as veracity. Big data's veracity is low and is said to be never 100% accurate, as there is lot of noise in this kind of data.

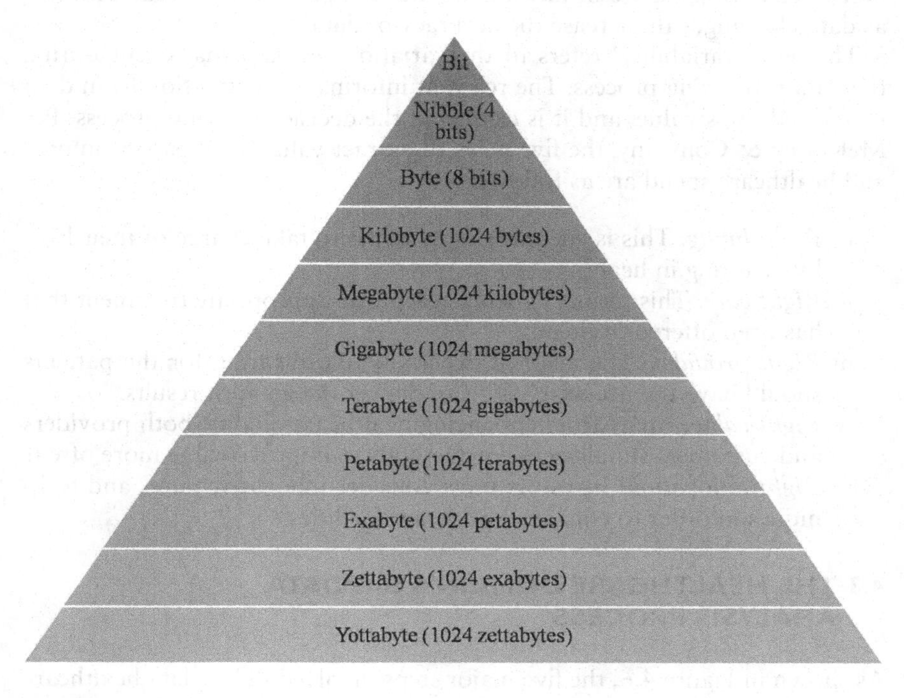

Figure 4.4 Data size pyramid.

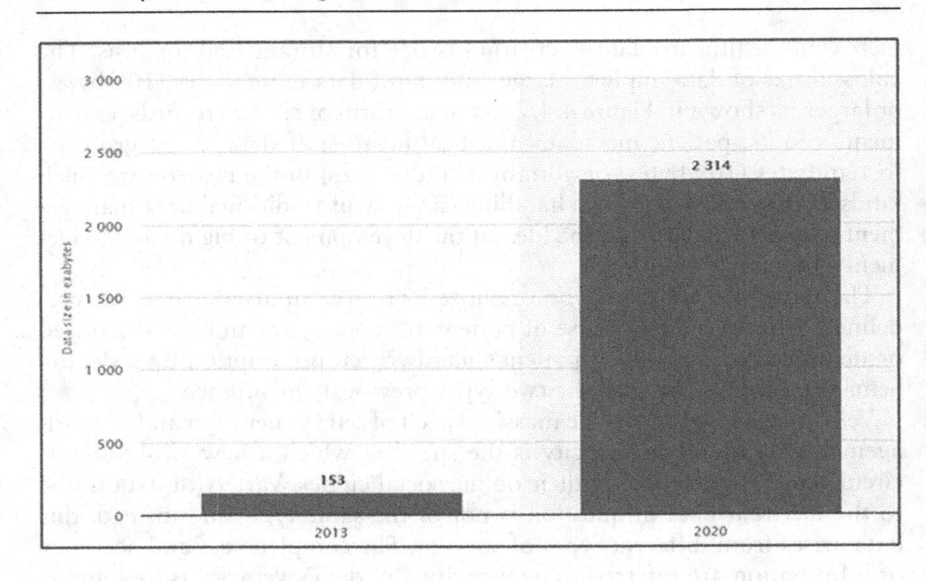

Figure 4.5 Global surge in healthcare data.

Noise refers to presence of unwanted data. Removal of noisy data, known as data cleaning, can increase the accuracy of data.

The term 'variability' refers to the variations in data that occur during the data processing process. The relevant information extraction from data is referred to as value, and it is critical in the decision-making process. Per McKinsey & Company, the five ways to extract value from patient impact and healthcare spend are as follows:

i *Right living:* This is encouraging patients to take charge of their lives by investing in health.
ii *Right care*: This means patients accept the appropriate treatment that has been offered to them.
iii *Right provider*: The medical experts who are caring for the patients should have the necessary skills to deliver the greatest results.
iv *Right value*: Apart from enhancing healthcare quality, both providers and customers should seek for methods to improve value more often.
v *Right innovation*: Investors must consider innovative cures and techniques in order to enhance healthcare facilities.

4.3 THE HEALTHCARE DOMAIN'S BIG DATA ANALYSIS PROCESS

As shown in Figure 4.6, the five major steps involved in big data healthcare analytics are data collection or acquisition, data storage, data management, data analytics and data visualization [21,24].

Figure 4.6 Major steps in big data analytics.

4.3.1 Data acquisition

Data acquisition is made up of three steps: data collection, data transmission and data pre-processing [25].

The healthcare data can be organized, semi-organized or unorganized, and can be acquired from primary sources or secondary sources. Sources of this data collection can be given as follows:

i. *Electronic health records (EHR):* The data from testing labs, reports of various medical tests, data captured by different types of sensors and doctor prescriptions all help in recommending personalized medicines to the patients and are a rich source of research studies for hospitals and research centres [26n].
ii. *Image processing:* Medical images from ultrasound, computed tomography (CT) scan, magnetic resonance imaging (MRI) and X-ray are some of the examples where image-related data is generated [12].
iii. *Social media:* Social media platforms like Facebook and Twitter may be used to collect public data on disease transmission and illness. This is achievable because individuals choose to use electronic media to spread illness information to their online circles.
iv. *Smartphones:* User health-related data can be gauged from various health monitoring apps installed in the user's mobile devices. AarogyaSetu App is one such app used by Indian citizens to track and monitor their symptoms related to COVID-19.

Sqoop, an open-source framework which works above the Hadoop Distributed File System (HDFS) ensures the exchange of data between HDFS and other relational databases [25]. Flume is used to aggregate and transfer data in chunks or batches from external systems to HDFS for storage

4.3.2 Data storage

Due to the ever-growing size of data, the storage of such data in databases with limited space is not feasible. Storage offered by big data framework and cloud storage are the most sought-after choices today, as data storage in the cloud drives the demand for scalable solutions [18]. Some of the renowned platforms that offer cloud storage are listed in Table 4.1:

Table 4.1 Cloud storage platforms

Data type	*Vendor*	*Platform*
Big data	Google	Google Cloud
	Amazon	Amazon Web Services
	Microsoft	Azure
	IBM	IBM Cloud

4.3.3 Data management

Data retrieval, data cleansing, missing data handling, data organization, data mining and data governance are all included in data management. In healthcare, data management aids in the risk assessment of patients and the provision of individualized treatments. Apache Ambari and HCatalog are two popular data management technologies. Data retrieval refers to the searching of medical text and medical images from a large collection of data. Data cleaning involves the removal of noisy data from the data of interest. In order to follow security and privacy norms, all patient-related data is governed by government acts, such as HIPAA, HITECH, HDI FOIA and GINA.

4.3.4 Data analysis

In this process, raw data is processed to yield useful information. Descriptive, diagnostic, predictive and prescriptive analytics are the four types of big data healthcare analytics. Descriptive analytics looks at past performance on historical data to determine the influence of a parameter on a system. Diagnostic analysis finds the root cause of a problem. Predictive analysis, upon looking at historical and real-time data, predicts the future. Prescriptive analysis is useful when a decision has to be made, as it suggests number of possible outcomes [28]. Pig is a tool designed to analyse large datasets and uses the programming language known as Pig Latin which is similar to SQL. Pig Latin scripts are converted into MapReduce processing tasks [25,29].

4.3.5 Data visualization

Graphical or pictorial representation of data for better understanding is known as data visualization. As complex data is difficult to understand without visualization, it acts as an aid to make better decisions.

4.4 BIG DATA APPLICATIONS IN THE HEALTHCARE DOMAIN

BD finds its applications in all healthcare areas, such as pharmaceutical development and personalized healthcare, insurance industry, and designing and manufacturing of medical devices [30].

4.4.1 Omics

To improve the precision of medical therapies, big data analysis may be applied to key organic and molecular databases known as omics. The study of genes is one of the omics known as genomics, and using big data analysis on it can assist to prevent and cure diseases as well as provide individualized treatment to patients. This has concentrated developments mainly on diseases like leukemia, cancer and diabetes. Study of proteome structure and functions, known as proteomics [31], will also find applications in big data by studying the protein structure and predicting and preventing human cancer. Prediction of epidemics based on geographical locations could be one unexplored area that leverages big data [32].

4.4.2 Insurance industry

Insurance industry uses big data analysis for fraud detection and claim management. This analysis could reveal if there are related claims by patient(s) being submitted to different insurance companies.

4.4.3 Pharmaceuticals

Big data is also used in pharmaceutical development. Drug discovery process is eased when using big data. Pfizer started a campaign to find ways to provide medications to patients more quickly.

4.4.4 Personalized patient care

Diagnostic assistance, medical prescriptions and clinical guidelines can be significantly improved with the usage of big data. The analysis of patient's healthcare data helps in providing online prescriptions and monitoring his/her symptoms online [25]. With the help of smartwatches worn by patients, real-time patient health monitoring has become possible which has opened the gateway of personalized recommendations to the patients based on his/her health choices [33,34].

4.4.5 Hospital management

It involves hospital management practices, such as preadmission testing, outpatient management and medication assignment. Cellular network optimization for outpatients might help in predicting medical conditions [32].

4.5 CHALLENGES OF BIG DATA ANALYSIS IN HEALTHCARE

The application of big data analysis (BDA) to the healthcare domain may face several challenges; some of which are listed subsequently [25,32,35,36]:

a. *Data quality:* The acquisition of big data may be affected if there's resistance from the organization to change their working model and pose as a challenge to the subsequent steps of big data analysis. Also, the collected data should be reliable, as this forms the major criterion for data selection during the data collection phase. Also, accuracy and consistency of the data should be ensured even after data pre-processing or cleaning is done.
b. *Initial investment:* Transitioning to BDA incurs huge initial costs to organizations providing healthcare.
c. *Privacy and security:* Healthcare or clinical data should be kept private and secure, as mining it can reveal patients' confidential data. Patients may have privacy and security concerns as their health-related information is vulnerable to unauthorized data access during intersystem exchanges [37].

4.6 CONCLUSION

Through this chapter, the implications of big data on healthcare were emphasized. Section 4.1 presented the concept of big data and healthcare. Its components and definition were discussed in Section 4.2. In Section 4.3, the process of big data was explained. Section 4.4 detailed the application of BD in healthcare. Based on this, one can contemplate the opportunities offered by BDA in healthcare. The analysis and insights gained from BDA in healthcare can help in making predictions, recommendations, diagnosis and personalized treatments. Although this is a positive outcome on the healthcare domain, the identified challenges should not be ignored for it to be the best possible choice for the healthcare domain.

REFERENCES

[1] Dash, S., S. K. Shakyawar, M. Sharma, and S. Kaushik, "Big data in healthcare: management, analysis and future prospects," *Journal of Big Data*, vol. 6, no. 1, pp. 1–25, Dec. 2019.
[2] Laney, D., "3D data management: controlling data volume, velocity, and variety | BibSonomy," *META GROUP*, 2001.
[3] Moore, K. D., K. Eyestone, and D. C. Coddington, "The big deal about big data," *Healthcare Financial Management*, vol. 67, pp. 60–68, 2013.
[4] Viceconti, M., P. Hunter, and R. Hose, "Big data, big knowledge: big data for personalized healthcare," *IEEE Journal of Biomedical and Health Informatics*, vol. 19, no. 4, pp. 1209–1215, Jul. 2015.
[5] Mehta, N., and A. Pandit. "Concurrence of big data analytics and healthcare: a systematic review," *International Journal of Medical Informatics*, vol. 114, pp. 57–65, 2018.
[6] Matheson, G. O., M. Klugl, and L. Engebretsen, "Prevention and management of non communicable disease: the IOC consensus statement," *Clinical Journal of Sport Medicine*, vol. 23, pp. 419–429, 2013.
[7] Wang, Y., et al. "Big data analytics: understanding its capabilities and potential benefits for healthcare organizations," *Technological Forecasting and Social Change*, vol. 126, pp. 3–13, 2018.
[8] Al-Jarrah, O. Y., P. D. Yoo, S. Muhaidat, G. K. Karagiannidis, and K. Taha, "Efficient machine learning for big data: a review," *Big Data Research*, vol. 2, no. 3, pp. 87–93, 1 Sept. 2015.
[9] Kuo, M.-H., et al. "Health big data analytics: current perspectives, challenges and potential solutions," *International Journal of Big Data Intelligence*, vol. 1, no. 1–2, pp. 114–126, 2014.
[10] Palanisamy, V., and R. Thirunavukarasu, "Implications of big data analytics in developing healthcare frameworks: a review," *Journal of King Saud University - Computer and Information Sciences*, vol. 31, no. 4, pp. 415–425, 1 Oct. 2019.
[11] Kumar, N. M. S., T. Eswari, P. Sampath, S. Lavanya, "Predictive methodology for diabetic data analysis in big data," *Procedia Computer Science*, vol. 50, pp. 203–208, 2015. doi:10.1016/j. procs.2015.04.069.
[12] van Horn, J. D., and A. W. Toga, "Human neuro imaging as a 'big data' science," *Brain Imaging and Behaviour*, vol. 8, pp. 323–331, 2014.
[13] Shilo, S., H. Rossman, and E. Segal, "Axes of a revolution: challenges and promises of big data in healthcare," *Nature Medicine*, vol. 26, no. 1, pp. 29–38, 1 Jan. 2020.
[14] Chen, M., Y. Hao, K. Hwang, L. Wang, and L. Wang, "Disease prediction by machine learning over big data from healthcare communities," *IEEE Access*, vol. 5, pp. 8869–8879, 2017.
[15] Pramanik, M. I., et al. "Healthcare informatics and analytics in big data," *Expert Systems With Applications*, vol. 152, p. 113388, 2020.
[16] De Mauro, A., M. Greco, and M. Grimaldi, "What is big data? A consensual definition and a review of key research topics," *AIP Conference Proceedings*, 2015, vol. 1644, no. 1, pp. 97–104.
[17] McAfee, A., and E. Brynjolfsson, "Big data: the management revolution," *Harvard Business Review*, vol. 90, no. 10, p. 4, Oct. 2012.

[18] Provost, F., and T. Fawcett, "Data science and its relationship to big data and data-driven decision making," *Big Data*, vol. 1, no. 1, pp. 51–59, Mar. 2013.

[19] Gandomi, A., M. Haider, "Beyond the hype: big data concepts, methods, and analytics," *International Journal of Information Management*, vol. 35, pp. 137–144, 2015.

[20] Sestino, A., M. I. Prete, L. Piper, and G. Guido, "Internet of things and big data as enablers for business digitalization strategies," *Technovation*, vol. 98, no. July, p. 102173, 2020.

[21] Feldman, B., E. M. Martin, T. Skotnes, "Big data in healthcare: hype and hope," *Dr. Bonnie 360 Degree* (Business Development for Digital Health), 2012. http://www.riss.kr/link?id=A99883549.

[22] Bello-Orgaz, G., J. J. Jung, and D. Camacho, "Social big data: recent achievements and new challenges," *Information Fusion*, vol. 28, pp. 45–59, 2016, ISSN 1566-2535, https://doi.org/10.1016/j.inffus.2015.08.005

[23] https://www.statista.com/statistics/1037970/global-healthcare-data-volume/

[24] Cottle, M., W. Hoover, S. Kanwal, M. Kohn, T. Strome, and N. Treister, *Transforming Health Care Through Big Data Strategies for leveraging big data in the health care industry* (Institute for Health Technology Transformation), 2013. http://ihealthtran. com/big-data-in-healthcare.

[25] Bahri, S., N. Zoghlami, M. Abed, and J. M. R. S. Tavares, "Big data for healthcare: a survey," *IEEE Access*, vol. 7, pp. 7397–7408, 2019.

[26] Bonney, S. "HIM's role in managing big data: turning data collected by an EHR into information," *Journal of American Health Information Management Association*, vol. 84–89, pp. 62–64, 2013.

[27] O'Driscoll, A., J. Daugelaite, and R. D. Sleator, "Bigdata, Hadoop and cloud computing in genomics," *Journal of Biomedical Informatics*, vol. 46, pp. 774–781, 2013.

[28] Dash, S., et al. "Big data in healthcare: management, analysis and future prospects," *Journal of Big Data*, vol. 6, no. 1, pp. 1–25, 2019.

[29] Tawalbeh, L. A., et al. "Mobile cloud computing model and big data analysis for healthcare applications," *IEEE Access*, vol. 4, pp. 6171–6180, 2016.

[30] Murdoch, T. B., and A. S. Detsky. "The inevitable application of big data to health care," *JAMA*, vol. 309, no. 13, pp. 1351–1352, 2013.

[31] Senthilkumar, S. A., B. K. Rai, A. A. Meshram, A. Gunasekaran, and S. Chandrakumarmangalam. "Big data in healthcare management: A review of literature," *American Journal of Theoretical and Applied Business*, vol. 4, no. 2, pp. 57–69, 2018.

[32] Khanra, S., A. Dhir, A. K. M. N. Islam, and M. Mäntymäki, "Big data analytics in healthcare: a systematic literature review," *Enterprise Information Systems*, vol. 14, no. 7, pp. 878–912, Aug. 2020.

[33] Farahani, B., et al. "Towards fog-driven IoT EHealth: promises and challenges of IoT in medicine and healthcare," *Future Generation Computer Systems*, vol. 78, pp. 659–676, 2018.

[34] Viceconti, M., et al. "Big data, big knowledge: big data for personalized healthcare," *IEEE Journal of Biomedical and Health Informatics*, vol. 19, no. 4, pp. 1209–1215, 2015.

[35] Huang, T., et al. "Promises and challenges of big data computing in health sciences," *Big Data Research*, vol. 2, no. 1, pp. 2–11, 2015.
[36] Kruse, C. S., et al. "Challenges and opportunities of big data in health care: a systematic review," *JMIR Medical Informatics*, vol. 4, no. 4, 2016.
[37] Abouelmehdi, K., et al. "Big healthcare data: preserving security and privacy," *Journal of Big Data*, vol. 5, no. 1, pp. 1–18, 2018.

Chapter 5

A novel scheme to manage the e-healthcare system using cloud computing and the internet of things

Nisha and Meenu Gupta

CONTENTS

5.1 Introduction 81
5.2 Background study 83
5.3 Internet of things (IoT) 86
5.3.1 Role of IoT in e-healthcare 86
5.3.2 Quality of healthcare using IoT 88
5.3.3 Novel authentication schemes for IoT-based healthcare systems 88
5.4 Cloud computing concept 88
5.4.1 Need for cloud computing 88
5.4.2 Mobile-based cloud computing for big data healthcare 89
5.4.3 Emergence of cloud in healthcare 90
5.4.4 Cloud role in e-healthcare system 92
5.5 Integration of IoT and cloud computing 92
5.6 Digital health evolution 93
5.7 Open issues and focusing factor 93
5.8 Tools for data analysis in e-healthcare 94
5.9 Digital imaging communication in medical and cloud computers 95
5.10 Conclusion and future scope 95
References 96

5.1 INTRODUCTION

The rapidly growing population has become a big challenge for people to live a better or healthy lifestyle. People face many problems regarding health and safety in life. To overcome such problems in today's scenario in

DOI: 10.1201/9781003305347-5

terms of health, several tools and technologies are on trend. In the case of the e-healthcare system, many mobile computing applications like oxygen saturation monitoring systems, glucose level capturing, body temperature sensing and m-healthcare, telemedicine, EMRs, PHRs mobile apps, information repositories, patient portals, and several other internet-based tools or software are available that are to be used. In addition, several portable devices and sensors are available that can measure heart rate (HR), blood pressure (BP), and respiration rate (RR) through a single touch. But still, there are some difficulties. The main problems arise in remote areas, the lack of computerized knowledge, and communications gap among doctors, patients, clinical staff and so others through the internet.

Figure 5.1 shows the relationship between the cost of care on the x-axis and the quality of life on the y-axis which explain how the quality of life can be improved by using the Internet of Things and the cost of care will be reduced [30]. In Figure 5.1, the quality of care vs. the cost of care are categorized into four parts: acute (or critical) care, residential care, home care, and health and wellness.

Figure 5.2 shows the acute care. Acute care is the short-term treatment/nursing for incident illnesses. It's a branch of the secondary healthcare

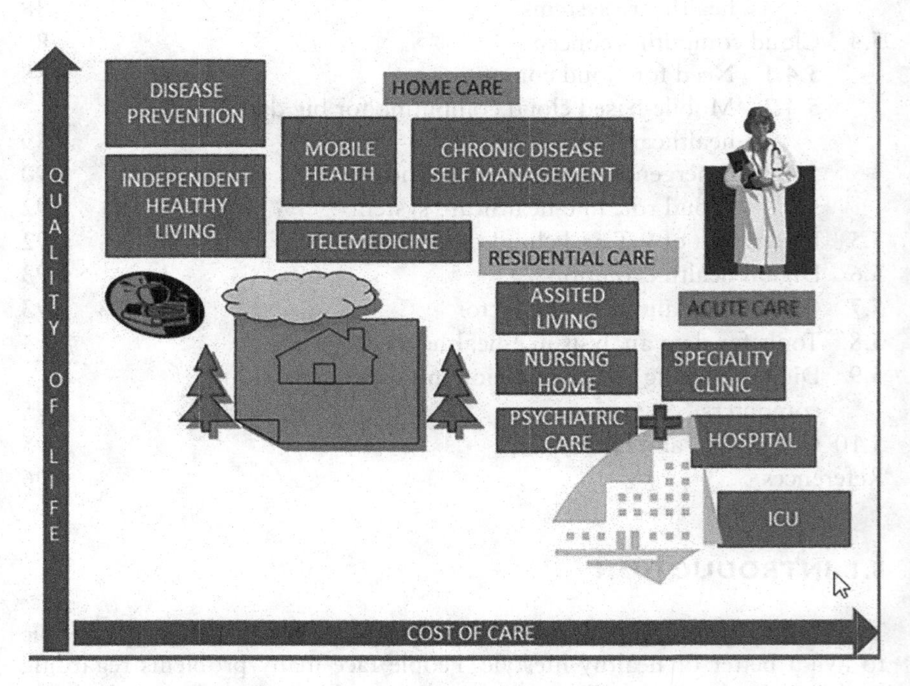

Figure 5.1 Relationship between the cost of care and quality of life.

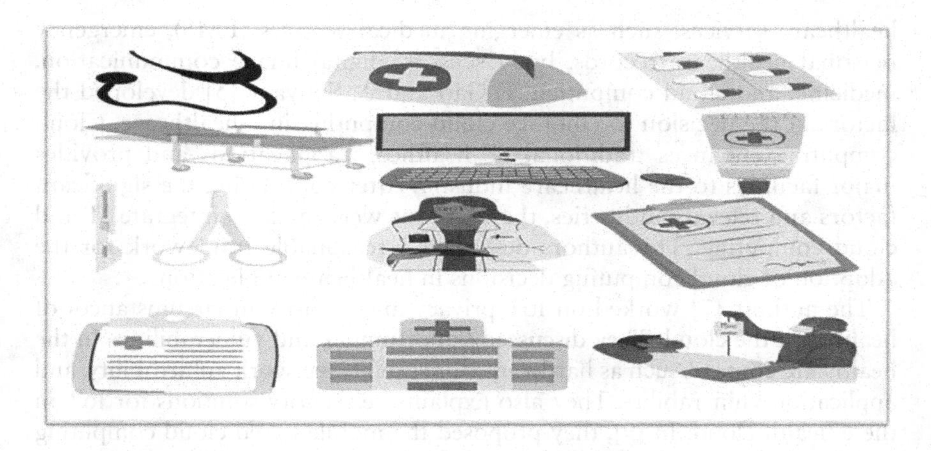

Figure 5.2 Critical (or acute) care.

system in which the victim has small-term treatments; for example, ICU, hospital, and special clinic.

Residential care is a special type of care for a person who can stay in a residential setting rather than their own home or family home. It is long-term care given to adults or children who need it; for example, assisted living and nursing homes.

Home care is a broad range of healthcare services that can be given in case of illness or injury. It is a more effective, convenient, and less expensive healthcare system that is the same as a hospital nursing facility (HNF); for example, mobile health, telemedicine, and chronic disease self-management.

Health and wellness is an online scheme that aims the promotion of e-healthcare awareness through education. It also provides a facility to collaborate with patients; for example, disease prevention and independent healthy living.

5.2 BACKGROUND STUDY

In [1], the authors discussed secure and compact communication on healthcare records using mobile-based cloud computing which includes the EHR access control schemes, fine-grained with linear secret-sharing schemes. L. Sun et al. [2] discussed e-healthcare with access control for cloud computing; they discussed the access control policies and connotation web. The goal of the connotation web is to automate the reuse, amalgamation, and communication of data among different web applications, and some overcome security issues. In [3], the authors analysed cloud computing and healthcare services, such as emerging medical services (EMS), emergency personal healthcare records, body sensors, digital image communication, medicine, and cloud computing. F. Gao and A. Sunyaev [5] developed the factors in the decision to embrace cloud computing in e-healthcare. Cloud computing enhances traditional IT healthcare applications and provides major facilities to the healthcare industry. After considering the significant factors and relevant industries, the decisions were made to integrate IT and cloud computing. The authors described a reasonable framework for the adoption of cloud computing decisions in healthcare applications.

The authors [7] worked on IoT privacy and security in circumstances of health and the cloud. They discussed opportunities and vulnerabilities in the healthcare context, such as hardware vulnerability, network vulnerability, and application vulnerability. They also explain the security solutions for IoT in the e-health cloud. In [9], they proposed the mobile-based cloud computing and big data scanning technique in healthcare applications. A large amount of data has been analysed to extract the value in terms of volume, velocity, variety, and veracity. The authors [10] discussed the validation artifices for IoT-based healthcare systems. They proposed that cyberattackers could attack computer hardware, software, and communication protocols for IoT ecosystems with enterprises, government systems, and industries. They also present the observations about security challenges, such as securing the IoT-networked devices by using secure cryptographic implementation and level of mutual connection between IoT-based resources. The use of a sensor tag mechanism can help identification at the same time and same place. In [11], they developed the importance of adopting the cloud computing model in the e-healthcare sector; they worked on current healthcare issues, the emergence of cloud in healthcare, the part of cloud computing in healthcare, and more.

In [12], the authors described the IoT-based e-healthcare systems that include ideas, expectations, and concerns, and also analysed the ideas and importance of IoT in the e-healthcare system. Further, in [14], the authors described the IoT-based distributed e-healthcare system. In DHS, all medical e-resources are interconnected to provide the facilities to everyone. In this work, the authors have also tried to enhance the interest of technical people, scientists, and researchers in the medical industry.

L. Rajbion et al. [16] analysed big data mechanisms and cloud computing roles. It has been used to store bulk amount of data on the cloud so that everyone can access their data quickly or easily through the internet. They also discussed some drawbacks and benefits of cloud computing and highlighted the use of big data in cloud computing in the future to provide an efficient healthcare system. In the present study, the authors analysed the various works on IoT and cloud computing where the reader can go through a novel scheme to manage the e-healthcare system by using IoT and cloud computing. A background study of different researchers is given in Table 5.1 and discussed the different aspects of smart healthcare systems using IoT and cloud computing technologies.

Table 5.1 Background study of various researchers regarding e-healthcare systems using IoT and cloud computing

Authors	*Year*	*Aspect discussed*	*Description*
Li., Y., Zh, Y., Lin, J., & Liu, Z. [1]	2018	Mobile cloud computing records, secure and fine-grained access control on e-healthcare	Secure and compact communication on healthcare records using mobile-based cloud computing which includes the EHR access control schemes, fine-grained with a linear secret-sharing scheme.
Aziz, H.A., & Guled, A. [3]	2016	Cloud computing and e-healthcare services	Cloud computing and healthcare services, such as EMS, emergency personal healthcare records, body sensors, digital image communication, medicine, and cloud computing.
Gao, F., & Sunyaev, A. [5]	2019	Context matters of determinant factors in deciding to adopt cloud computing in e-healthcare	Cloud computing enhances traditional IT healthcare applications and provides major facilities to the healthcare industry. After considering the significant factors and relevant industries, the decisions were made to integrate IT and cloud computing. The authors proposed a conceptual framework for adopting cloud computing decisions in e-healthcare applications.
Lo., A.T., M, R., Benkhlifa, E., & Song, H. [9]	2016	Big data, analysis based on mobile cloud computing model for smart healthcare applications	Big data analysis technology has been used to extract the value in terms of volume, velocity, variety, and veracity.
Hou, J. L., & Yeh, K. H. [10]	2015	Novel authentication schemes in healthcare systems using IoT	The authors discussed cyberattackers that could attack computer hardware, software, and communication protocols for IoT ecosystems with enterprises, government systems, and industries. They also presented the observations about security challenges, such as securing the IoT-networked devices by using secure cryptographic implementation and level of mutual connection between IoT-based resources. The use of a sensor tag mechanism can help identification at the same time and same place.

(*Continued*)

Table 5.1 (Continued)

Authors	*Year*	*Aspect discussed*	*Description*
Bami., M., Brohi, S., & Chuprat, S. [11]	2012	Adopting a cloud computing paradigm on the significance of the e-healthcare sector	Authors worked on current healthcare issues, the emergence of cloud in e-healthcare, cloud computing's role in e-healthcare, and more.
Maksimović, M., & Vujović, V. [12]	2017	Ideas, expectations, and concerns with the Internet of Things–based e-healthcare	Internet of things (IoT)–based healthcare includes ideas, expectations, and concerns. And they also analysed the ideas and importance of IoT in the e-healthcare system.

5.3 INTERNET OF THINGS (IOT)

IoT suggests that the Internet of Things is a collection of individuals' computer systems/objects through the internet. IoT is a device that connects everyone wirelessly through the internet and can transfer the data of patients. In other words, the connection of all those things, such as electronic components, software, and sensors, to gather and interchange data without human interference is called the Internet of Things (IoT). In the e-healthcare system, the area of IoT lets the patient take online appointments and disease consultations on behalf of a doctor from a facility. The doctor can take an online record of a patient's disease, and that will be used for future references, and cloud computing in e-healthcare can be used to store and handled a large number of patient's disease data on the cloud and mainly on central-level servers so both doctors and patients can refer to this data.

5.3.1 Role of IoT in e-healthcare

IoT devices in e-healthcare play a vital role by reducing costs, being user friendly, improving the quality of life of patients, and so on. There are many reasons that healthcare applications move towards IoT devices. The role of IoT has to be increased in many areas, such as maintaining vital equipment, monitoring medical assets, monitoring patient care, and tracking equipment usage. The physician can monitor the patient's health condition remotely.

IoT platform: It is an integrated service that bridges the gap between the IoT devices and the healthcare system and offers physical objects online. Some IoT platforms are as follows:

- Amazon Web Services (AWS)
- Google Cloud

- IBM Watson
- Artik Cloud

IoT platforms are helpful in collecting visualized data gathered by the sensor and device:

- It connects the sensor and devices.
- It handles different software communication protocols and hardware.
- It provides security and authentication for sensors and users.

IoT, through an e-healthcare system, fetches several facilities, including savings, transportation, healthcare, and insurance costs. It also improves the quality of care in e-healthcare systems [10]. Figure 5.3 shows the benefits of an IoT-cloud-based e-healthcare system that includes cost reduction, accessibility, online assistance, big data processing, efficient healthcare, and ease of use. Furthermore, the utilization of IoT services in electronic healthcare systems has increased concerns in privacy and security, but still, several issues also arise here, like sensitive data in electronic medical records being

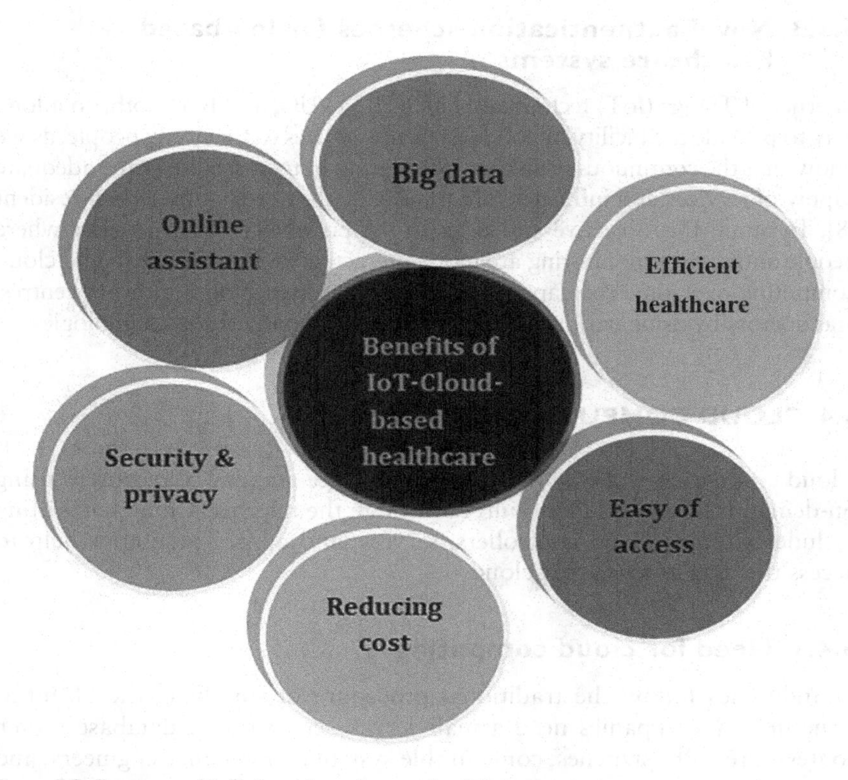

Figure 5.3 Features of IoT-cloud-based smart healthcare systems.

accessed by multiple points, and data management is also a key issue in IoT because of interconnected devices exchanging a large quantity of data that are necessity to be stored and processed correctly.

5.3.2 Quality of healthcare using IoT

There are mainly three qualities that validate the sensor-equipped thing to become a part of IoT healthcare:

- Firstly, the device should be hearing and collecting the data about the surroundings, such as temperature, humidity, blood oxygen, electrocardiogram, and blood glucose.
- Secondly, the devices should be automatically collecting the data at the central level as a central coordinator (with other devices also or in case of any emergencies).
- Finally, it should be running until the process is completed; for example, if a patient's blood sugar level or blood pressure level is high in that critical situation, a high alert should be triggered for immediate action.

5.3.3 Novel authentication schemes for IoT-based healthcare systems

Internet of Things (IoT) techniques, like ICT, 3G/UG, and Bluetooth, go a long way to provide the facility of good (efficient) services to the needy people, as we know that the continuous growth of population in towns calls for the adequate supply of services and infrastructure to provide the needs of the urban resident [8]. To attain this objective and react to the possibility of smart cities where people and service monitoring are more interactive and efficient, IoT and cloud computing overcome the gap among the people (patients), healthcare centres, and doctors by using various information and communication technologies.

5.4 CLOUD COMPUTING CONCEPT

Cloud computing is a way that provides a huge amount of resources using on-demand delivery of IT resources through the internet. Cloud computing includes storage, cloud controllers, servers, and apps. These apps help to access the datastore on that cloud.

5.4.1 Need for cloud computing

IT industries follow the traditional procedure of supplying the IT infrastructure. IT companies need a mail server, server room, database server, routers, firewalls, switches, configurable system maintenance engineers, and high-speed data network; to set up such IT infrastructures, they require a lot of finances [30]. To handle all these types of trouble and to minimize the IT infrastructure cost, cloud computing comes into trend [23].

Figure 5.4 shows the imaginary skeleton of the determinant factor for the adoption of cloud computing in e-healthcare.

The dotted line rectangle represents the technologies, environment, and organization which are useful for describing the adoption of cloud computing in e-healthcare.

The solid line rectangle represents the data/information and stakeholders. Data is the unsymmetrical manner that is serviced as a relevant conceptual category, and also only limited stakeholder roles have been examined [5].

The arrow represents the dual role of the industry in the identification of relevant category concepts; specific variables are generalized by the public role of healthcare and clinical role.

5.4.2 Mobile-based cloud computing for big data healthcare

In a traditional infrastructure of mobile cloud computing, the user accesses the cloud resources remotely through the internet, but in recent scenarios, big data is associated with a massive amount of stored data due to revolutionary enhancements in different technologies, such as wireless communication technology, cloud computing, and the spread in social media [22]. Big data infrastructure depends upon the size of data, time frequency, and type of data – every minute, day, month, or decade. Big data in e-healthcare is useful to execute real-time tasks to analyse the data and extract the useful data from medical images. These help to make the right decisions between patient life and death.

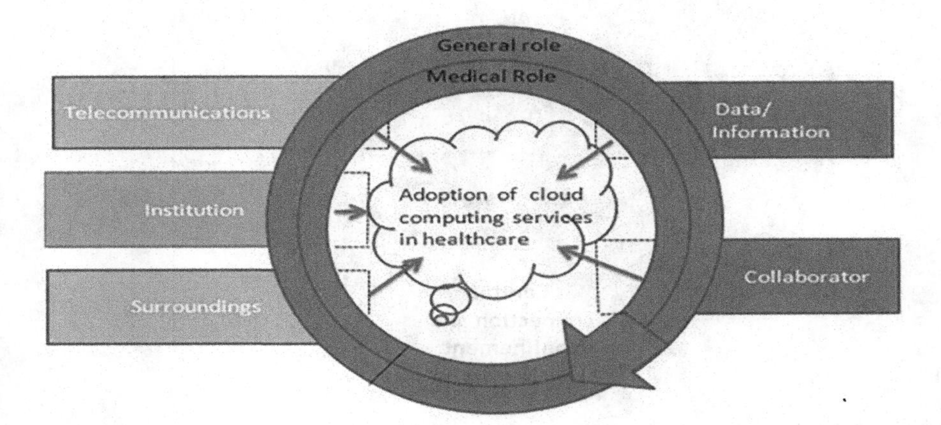

Figure 5.4 Imaginary skeleton of determinant parts affecting cloud computing in e-healthcare.

5.4.3 Emergence of cloud in healthcare

It provides unique facilities, including on-demand-based self-services, emergence network access, payment according to self-requirement/use pattern, rapid flexibility, and resource pooling like that [31]. Many offers for a variety of service delivery, such as SaaS, IaaS, and PaaS, are also coming into the trend of cloud in e-healthcare [19].

Cloud service models are as follows:

- Software as a service (SaaS): based on on-demand software
- Infrastructure as a service (IaaS): maintains IT infrastructures
- Platform as a service (PaaS): helps developers

Figure 5.5 shows the three main model services in cloud computing, that is SaaS, IaaS, and PaaS. SaaS is software distributed representation in which an application is controlled by a cloud service supplier and constructed and accessible to customers through the internet. It is the user-level applications that provide the platform to the user and allow the

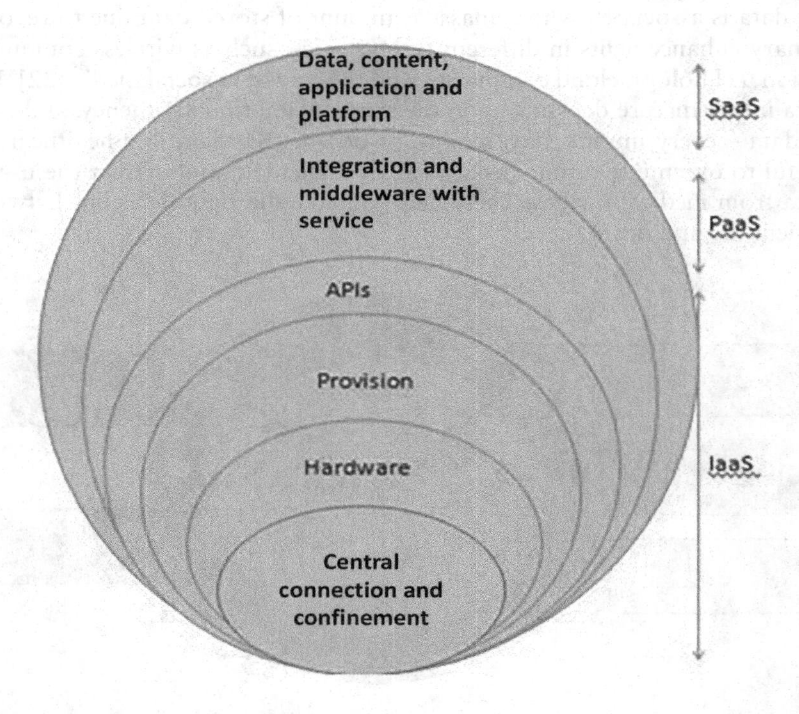

Figure 5.5 The functions of the cloud services model.

entering of data or content on that level. PaaS is the intermediate part of the SaaS and IaaS that includes the integration and middleware with the services. The IaaS is used to do the core connectivity and delivery facility through APIs.

- IaaS helps customers scale the arrangement to meet exchanging needs and announce the services used. IaaS is offered in three models, such as private, public, and hybrid.
- The private cloud has been used for a single organization.
- The public cloud is used for general industry or public groups.
- A hybrid cloud has combined the characteristics of one or more cloud computing models.

Hence, the selection of these distributed services models and cloud delivery depends upon the requirements of management and the client's data sensitivity which is shown in Figure 5.6 clearly.

- PaaS is that type of service that allows the developer to develop a web application without installing any software or tool on the computer. These are hosted in the cloud and accessed through the browser.

The cloud computing delivery services are as shown in Figure 5.6:

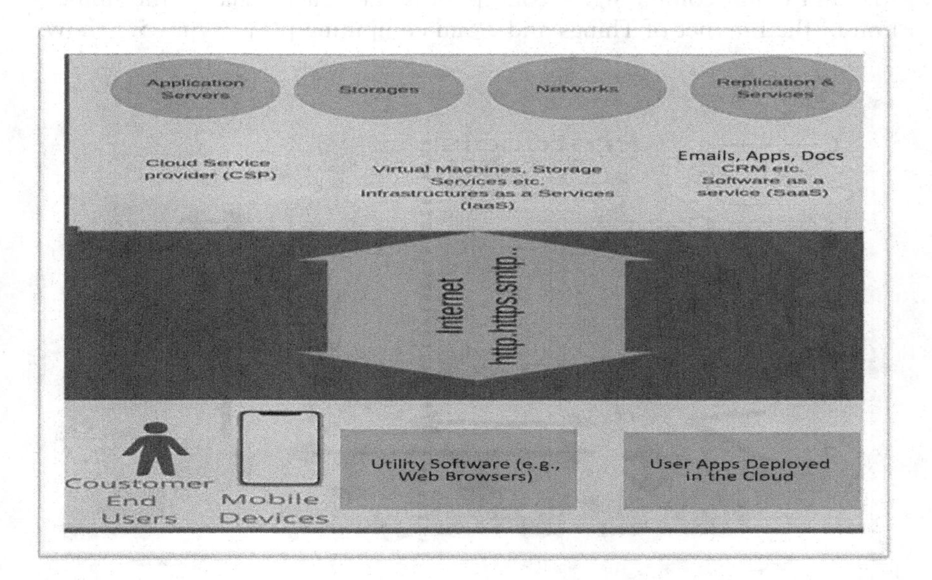

Figure 5.6 Cloud computing architecture.

5.4.4 Cloud role in e-healthcare system

Well, cloud computing is used in many sectors, but if we talk about healthcare, it acts as a link between the doctor and patient [19]. It helps old-age people and rural and urban people that are living far from hospitals and cannot reach it at the right time in difficult situations.

If we move forward to the next generation of healthcare systems, it involves smart home delivery, smart emergency hospitals, smart control devices, and so on that are interconnected with each other; for example, IBM companies are merged with several industries to work together to solve various issues of technology [24]. Cloud computing helps to immediately access patient's lab test output and further allows physicians to check those results immediately. The main role of cloud computing is to provide a large amount of data space on the cloud so that everybody can access these data very easily and quickly as soon as possible by using a smart healthcare system. It operates on the network to track the patient information which helps medical investigation and practices as a fast response for diseases, bio-surveillance for biological or chemical attacks, and so on [23].

5.5 INTEGRATION OF IOT AND CLOUD COMPUTING

Insufficient use of IoT resources may cause the cloud computing environment to allow the sending of incorrect data. So to overcome this gap between IoT and cloud computing, a concept was introduced that is the unification of the Internet of Things and cloud computing [21]. Figure 5.7 shows

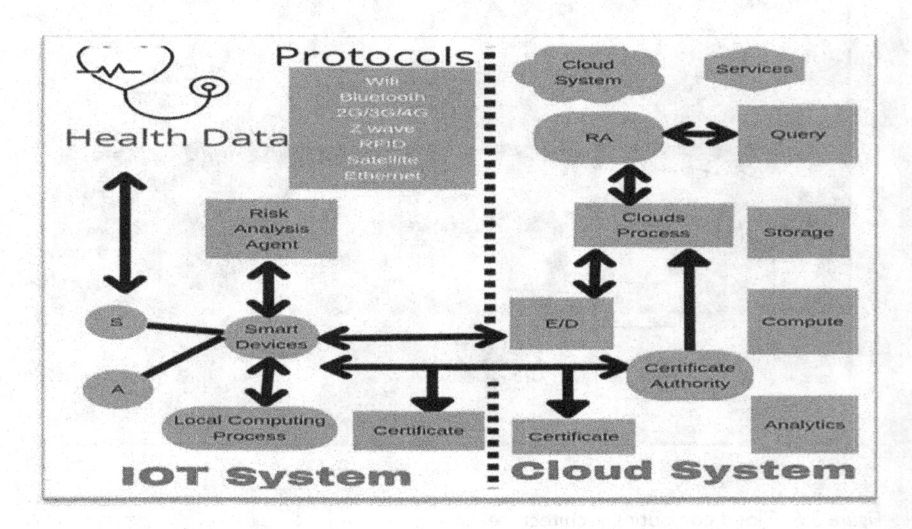

Figure 5.7 IoT and cloud computing.

the distribution of security and processing of health information. By using smart technologies, resources not only assemble the data but also insert privacy methods and encrypt data before sending it to the cloud environment [7]. The author proposed a temporary backup in the IoT environment to ensure the presence of data in case of network error [29]. And also, a certificate authority (CA) was included to protect the exchange of information. Authors also added standard methods of access control and cryptography to guess and protect the risk of damages.

5.6 DIGITAL HEALTH EVOLUTION

The basic idea is to connect the different types of objects, like sensors, cell phones, and radio-frequency identification (RFID) targets. Therefore, a lot becomes virtual, meaning everything is easily locatable, readable, and addressable on the network. M-healthcare applications provide some of the most cloud computing and IoT technologies, like disturbed healthcare system and telemedicine, and it has an IoT and cloud computing–based healthcare system [20].

It reflects multiple provisions, such as remote monitoring of continuous data, so that patients can be monitored day by day with sensors in mobile devices [28]. In the 1970s, the use of the digital system was included in accounting, payroll, invoices, and census. The mobile phone, digital images, DICOM, disparate databases, minimal datasets, smart cards, and other facilities came in 1990s. In the 2000s era, more development has been made, such as the RFID, linear bar code, robotics, and mobile computing. Now, with continuous enhancement of new tools and technologies, like IoT, cloud computing, sensor networks, 5G, self-management of personalized healthcare, teleportation, biometrics, and quantum inscription, there are so many advanced tools and technologies that are used and will be used in the future of the healthcare system [25]. Several services that have been used in e-healthcare are in the public healthcare and personal healthcare systems. These are mainly two services, including e-healthcare public healthcare services, considering guidelines regarding drugs and food, and sustaining to maintain a healthy environment; whereas personal healthcare services are supplied at home, in hospitals, and in relevant organizations [4]. Teaching and research activities are also an essential part of the prevention, tracking, detection, and treatment of diseases. The advanced innovations of smart electronic devices help with the dynamic nature of healthcare delivery [15]. The success of healthcare will depend on the communication among users, processes, data, rules/policies, and procedures.

5.7 OPEN ISSUES AND FOCUSING FACTOR

IoT brings a wide arrangement of advanced security issues for the research group in common system security, nature, and application security because cyberattackers normally use security vulnerabilities in computers hardware/

software and communication protocols to target the IoT ecosystem within enterprises, institutes, and government systems [18]. So there is a need to focus on confidentiality, probity, and availability of these determined concepts. Figure 5.8 shows various challenges and open issues in IoT and cloud computing.

So there is a need for the alteration of technical-based solutions to handle the challenges in medical environments.

5.8 TOOLS FOR DATA ANALYSIS IN E-HEALTHCARE

Data analysis means examining the raw material/data to clean, inspect/conclude, and model and convert these data into useful information. There are various tools for data analysis: Data Wrangler is useful for cleaning and rearranging data. Google Refine uses a spreadsheet to handle the text or numeric data. The R Project is run on the command line; it is also called a statistical computing device. It is also used to make a chart or graph and plot the results. Google Fusion Tables is also used to create a chart or map. Other tools are Tableau Public, OpenStreetMap time flow, IBM Word Cloud Generator, JavaScript InfoVis Toolkit, QGIS, and Talend Open Studio for Big Data – all these are tools that are used for data analysis.

There are mainly three qualities that validate the sensor-equipped thing to become a part of IoT healthcare. Firstly, the device should be hearing and collecting the data about the surroundings, such as temperature, humidity, blood oxygen, electrocardiogram, and blood glucose. Secondly, the devices should automatically collect the data at a central level as a central

challenges and issues

infrastructure standardization IoT healthcare plateform advanced hardware design software architecture quality of service mobility network type energy saving cost analysis etc.	data management dat analysis integration data protection data representation data deposition etc.	figuring intelligence advanced method calculation such as edge/fog computing IoT-DHS etc.

Figure 5.8 Categorization of open issues.

coordinator (with other devices also or in case of any emergencies). Finally, it should be running until the process is completed; for example, if a patient's blood sugar level or blood pressure level is high in that critical situation, a high alert should be triggered for immediate action [27].

5.9 DIGITAL IMAGING COMMUNICATION IN MEDICAL AND CLOUD COMPUTERS

Cloud computing plays a central role in storing digital medical images, and after that, the images have to be accessed by using PACS, DICOM, or others in emergencies. PACS is a tool that stands for picture archiving and communication system (PACS) which has been used to store medical images. This tool is used in clinical settings to handle all medical images that are useful for clinical support. PACS helps to store, handle, retrieve, and distribute the medical images. Here, critical support means that when using these medical images, decisions and difficult treatment strategies have to be taken. Another tool is DICOM (Digital Imaging and Communications in Medicine) which supports storage formats and complex medical images. But the use of DICOM is restricted within virtual private network (VPN) connections. Web 2.0 is partially well suited with DICOM, and it only allows image retrieval over the web. These restrictions avert medical staff in external institutions from accessing the PACS archive. Medical organizations depend on these processes, including VPNs that required point-to-point configuration and are not scalable. However, this method is applicable only when cloud computing services are integrated with DICOM to solve emergencies when medical images are needed to be accessed remotely. With the help of cloud computing, medical images are accessed directly from EHR or PHR with the use of mobile devices.

5.10 CONCLUSION AND FUTURE SCOPE

In this chapter, the use and role of IoT and cloud computing in e-healthcare systems are discussed that make life easier and provide very quick results with minimum cost facility to everyone, even those living at home. IoT infrastructure just provides the availability of IT resources, but cloud computing, with the storage of big data, also provides the facility to access the better delivery of services. Cloud computing in the healthcare industry focuses on increasing the efficiency of delivered healthcare services rather than their own IT infrastructure management. The main aim of cloud computing integration with IoT is to provide scalable communication and on-demand delivery of services without the need to purchase new hardware and software. There is also some security and privacy issues and challenges. Using some advanced tools and technology, we provide a better platform in the e-healthcare system at a low cost.

REFERENCES

[1] Liu, Y., Zhang, Y., Ling, J., & Liu, Z. (2018). Secure and fine-grained access control on e-healthcare records in mobile cloud computing. *Future Generation Computer Systems*, *78*, 1020–1026.

[2] Sun, L., Wang, H., Yong, J., & Wu, G. (2012). Semantic access control for cloud computing based on e-healthcare. In *Proceedings of the 2012 IEEE 16th International Conference on Computer Supported Cooperative Work in Design (CSCWD)*, Wuhan (pp. 512–518). doi:10.1109/CSCWD.2012.6221866.

[3] Aziz, H. A. (2015). Health informatics: Introduction. *Clinical Laboratory Science*, *28*, 238–239.

[4] Aziz, H. A., & Guled, A. (2016). Cloud computing and healthcare services. *Journal of Biosensors & Bioelectronics*, *7*(3). doi:10.4172/2155-6210.1000220.

[5] Gao, F., & Sunyaev, A. (2019). Context matters: A review of the determinant factors in the decision to adopt cloud computing in healthcare. *International Journal of Information Management*, *48*, 120–138.

[6] Ammenwerth, E., Iller, C., & Mahler, C. (2006). IT-adoption and the interaction of task technology and individuals: A fit framework and a case study. *BMC Medical Informatics and Decision Making*, 6(3), 479. doi:10.1186/1472-6947-6-3.

[7] Ida, I. B., Jemai, A., & Loukil, A. (2016, December). A survey on security of IoT in the context of eHealth and clouds. In *2016 11th International Design & Test Symposium (IDT)* (pp. 25–30). IEEE.

[8] Hou, J.-L., & Yeh, K.-H. (2015). Novel authentication schemes for IoT based healthcare systems. Hindawi Publishing Corporation, *International Journal of Distributed Sensor Networks*, *2015*, 9, Article ID 183659.

[9] Lo'ai, A. T., Mehmood, R., Benkhlifa, E., & Song, H. (2016). Mobile cloud computing model and big data analysis for healthcare applications. *IEEE Access*, *4*, 6171–6180.

[10] Hou, J. L., & Yeh, K. H. (2015). Novel authentication schemes for IoT based healthcare systems. *International Journal of Distributed Sensor Networks*, *11*(11), 183659.

[11] Bamiah, M., Brohi, S., & Chuprat, S. (2012, December). A study on significance of adopting cloud computing paradigm in healthcare sector. In *2012 International Conference on Cloud Computing Technologies, Applications and Management (ICCCTAM)* (pp. 65–68). IEEE.

[12] Maksimović, M., & Vujović, V. (2017). Internet of Things based e-health systems: Ideas, expectations and concerns. In *Handbook of Large-Scale Distributed Computing in Smart Healthcare* (pp. 241–280). Springer, Cham.

[13] Birje, M. N., & Hanji, S. S. (2020). Internet of things based distributed healthcare systems: a review. *Journal of Data, Information and Management*, 2(3), 149–165.

[14] Hou, J. L., & Yeh, K. H. (2015). Novel authentication schemes for IoT based healthcare systems. *International Journal of Distributed Sensor Networks*, *11*(11), 183659

[15] Raj, C., Jain, C., & Arif, W. (2017, March). HEMAN: Health monitoring and nous: An IoT based e-health care system for remote telemedicine. In *2017 International Conference on Wireless Communications, Signal Processing and Networking (WiSPNET)* (pp. 2115–2119). IEEE.

[16] Rajabion, L., Shaltooki, A. A., Taghikhah, M., Ghasemi, A., & Badfar, A. (2019). Healthcare big data processing mechanisms: The role of cloud computing. *International Journal of Information Management*, *49*, 271–289.
[17] Ahuja, S. P., Mani, S., & Zambrano, J. (2012). A survey of the state of cloud computing in healthcare. *Network and Communication Technologies*, *1*(2), 12.
[18] Butpheng, C., Yeh, K. H., & Xiong, H. (2020). Security and privacy in IoT-cloud-based e-health systems: A comprehensive review. *Symmetry*, *12*(7), 1191.
[19] Abbas, A., Bilal, K., Zhang, L., & Khan, S. U. (2015). A cloud based health insurance plan recommendation system: A user centered approach. *Future Generation Computer Systems*, *43*, 99–109.
[20] Yu, L., Lu, Y., & Zhu, X. (2012). Smart hospital based on internet of things. *Journal of Network*, 7, 1654–1661.
[21] Doukas, C., & Maglogiannis, I. (2012). Bringing IoT and cloud computing towards Pervasive Healthcare. In *Sixth International Conference on Innovative Mobile and Internet Services in Ubiquitous Computing, Palermo* (pp. 922–926).
[22] Jara, A. J., Zamora, M. A., & Skarmeta, A. F. (2012, June). Knowledge acquisition and management architecture for mobile and personal health environments based on the internet of things. In *2012 IEEE 11th International Conference on Trust, Security and Privacy in Computing and Communications* (pp. 1811–1818). IEEE.
[23] Srinivasan, M. et al. (2012). State-of-the-art cloud computing security taxonomies: A classification of security challenges in the present cloud computing environment. In *International Conference on Advances in Computing, Communications and Informatics (lCACCi-20i2), ICACCI'12, ACM, 2012, Chennai, India.*
[24] Adebesin, F., Kotzé, P., Van Greunen, D., & Foster, R. (2013a). *Barriers and challenges to the adoption of E-Health standards in Africa.* Health Informatics South Africa (HISA), Port Elizabeth, 3–5 July 2013, http://hdl.handle.net/10204/6910.
[25] Fricker, S. A. et al. (eds.) (2015). *Requirements Engineering for Digital Health.* Springer International Publishing, Switzerland.
[26] MacIntosh, E., Rajakulendran, N., & Salah, H. (2014). Transforming health: Towards decentralized and connected care. *MaRS Market Insights.* https://www.marsdd.com/news/transforming-health-decentralized-connected-care/
[27] Berhanu, Y., Abie, H., & Hamdi, M. (2013). A testbed for adaptive security for IoT in eHealth. In *Proceedings of the International Workshop on Adaptive Security (ASPI '13)*, article 5.
[28] Tozlu, S., Senel, M., Mao, W., & Keshavarzian, A. (2012). Wi-Fi enabled sensors for internet of things: A practical approach. *IEEE Communications Magazine*, *50*(6), 134–143.
[29] Chen, D., Chang, G., Jin, L., Ren, X., Li, J., & Li, F. (2011). A novel secure architecture for the Internet of things. In *Proceedings of the 5th International Conference on Genetic and Evolutionary Computing (ICGEC '11)* (pp. 311–314). IEEE, Xiamen, China, September 2011.
[30] Zissis, D., & Lekkas, D. (2012). Addressing cloud computing security issues. *Future Generation Computer Systems*, *28*, 583–592.
[31] Koufi, V., et al. (2010). Ubiquitous access to cloud emergency medical services. In *10th International Conference on Information Technology and Applications in Biomedicine (ITAB)* (pp. 1–4). IEEE.

Chapter 6

Automating remote point-of-care ECG diagnostics via decentralized report routing algorithm

Bidyut Bikash Borah, Satyabrat Malla Bujar Baruah, Debaraj Kakati, and Soumik Roy

CONTENTS

6.1 Introduction 99
6.2 Methods 100
6.2.1 ECG acquisition system and local server 101
6.2.2 Decentralized data routing 103
6.2.3 Routing algorithm at node 103
6.3 Results and discussion 105
6.4 Conclusion 109
References 112

6.1 INTRODUCTION

In recent years, most developing countries have not had enough medical care facilities to diagnose the expanding number of patients. Despite having a modern and well-equipped medical facility, the sector suffers from a shortage of experts and human resources. The Covid-19 pandemic's recent improvements have added to the present medical scenario and diagnoses [1],[2]. Such problems cannot be solved by combining high-tech facilities and sophisticated tools but rather by enlisting the early help of specialists to conduct proper evaluations [3],[4],[5]. The cardiovascular health of patients is one such essential diagnosis that necessitates rapid evaluation by cardiologists [6],[7],[8]. According to the survey, the majority of cardiovascular deaths are caused by delayed diagnosis and lack of prompt medical attention. However, due to various factors, such as a scarcity of cardiologists, inadequate diagnostic equipment, and patients' lack of access to specialists, the diagnosis and medical consultation process suffer. The introduction of effective and speedy diagnostic facilities, such as telemedicine and e-healthcare services that allow for remote assessment of patient's health, is an easy method to reduce such losses [9],[10],[11]. Current e-healthcare and

DOI: 10.1201/9781003305347-6

telemedicine trends include human resource intervention in routing critical records between specialists and remote nodes, which suffers significantly as the number of patients grows [12],[13]. The GE MAC-600 is one such system that can collect patients' ECGs from distant locations and send the data to specialists for further evaluation by the nodal administrator [14]. An in-house patient monitoring system with computational intelligence can solve these issues. In remote healthcare, an in-house patient monitoring system gives acceptable results and more adaptability [15],[16],[17].

A novel intelligent medical report route computing algorithm has been designed and implemented in this proposed work to eliminate human intervention and improve ECG diagnostic response time. For the acquisition of patient ECG signals, a server-client configuration has been set up. This acquisition system incorporates an intelligent report routing mechanism that sends patients' ECG data to remote specialists for evaluation right away. The system prioritizes quick examination and routes ECG data based on cardiologists' essential factors, such as reaction time, efficiency, and cooling periods, and updates the same depending on their responses. Additionally, its mobility and real-time transmission capabilities, simplicity of use, and low cost make it an attractive alternative for prehospital cardiac care in clinical and community settings. Because any caregiver or medical personnel efficiently operates the suggested system, it is also beneficial for physicians who desire to monitor their patients remotely.

6.2 METHODS

The proposed ECG acquisition system is shown in Figure 6.1, equipped with the medical data routing algorithm for fast-track diagnosis and report generation. The proposed system acquires ECG data from patients in remote locations with metadata (patient name, age, unique identifier, measurement location, etc.) for patient identification. It transmits the acquired data to a

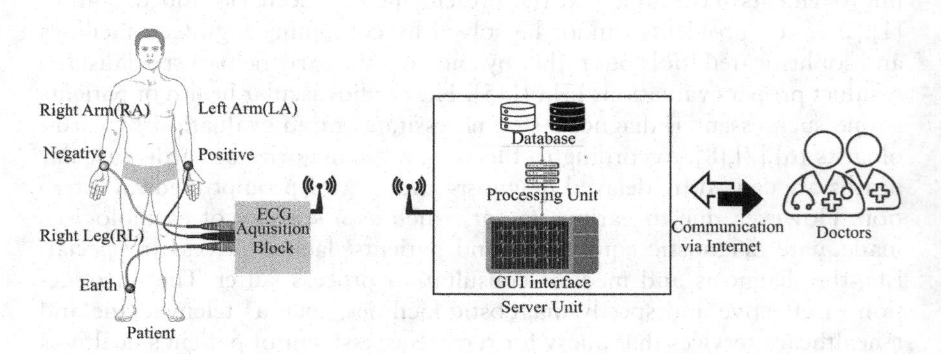

Figure 6.1 ECG acquisition system equipped with an auto-routing algorithm.

local server (server unit) via a Wi-Fi module for diagnostics and future reference. The server also facilitates a user prompt to enter additional data, such as medical history, mode of report reception (via e-mail or in-person) with additional fields for e-mail, gender, diabetic status, and other medical conditions. The server is responsible for decentralized local storage and transmission of the undiagnosed patient's data to registered doctors for immediate assessment. Diagnosis reports of patients are automatically generated after receiving cardiologists' reviews. The details of the system are discussed in the sections following.

6.2.1 ECG acquisition system and local server

Shown in Figure 6.2 is the schematic of the ECG acquisition system responsible for transducing the electrocardiogram signal. The AD8232 integrated chip has been used in the analog front end (AFE) for ECG acquisition of the proposed portable ECG diagnostics system. The main feature is a one-of-a-kind input stage architecture: a transconjunctival INA with a high CMRR based on two highly complementary current amplifiers (up to 80 dB). A clean signal may be produced without any additional notching filters. The built-in integrator minimizes DC offset drift caused by defective lead contact by serving as a second-order HPF. The installation of an operational amplifier enables the use of the inbuilt LPF, which provides the required gain value.

The acquisition system (Figure 6.3) is connected to a decentralized server via a Wi-Fi module (ESP8266) for local storage and auto-routing of the acquired

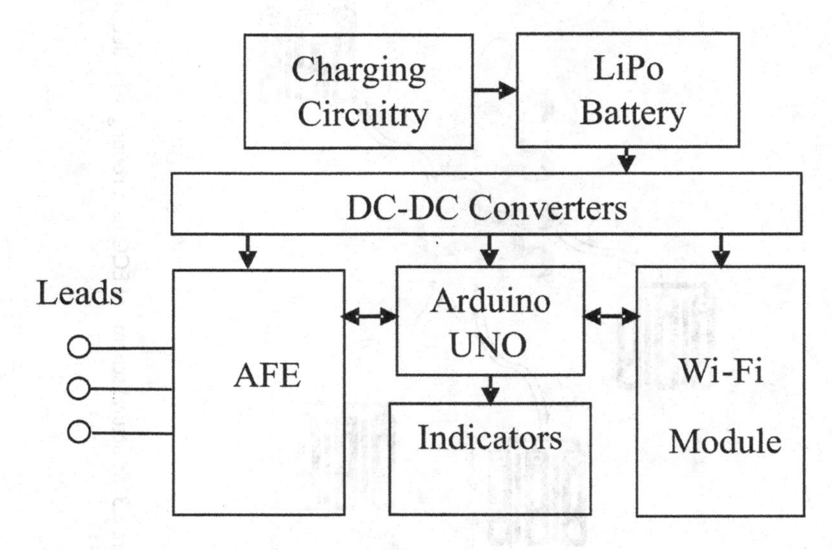

Figure 6.2 Structural scheme of ECG acquisition system.

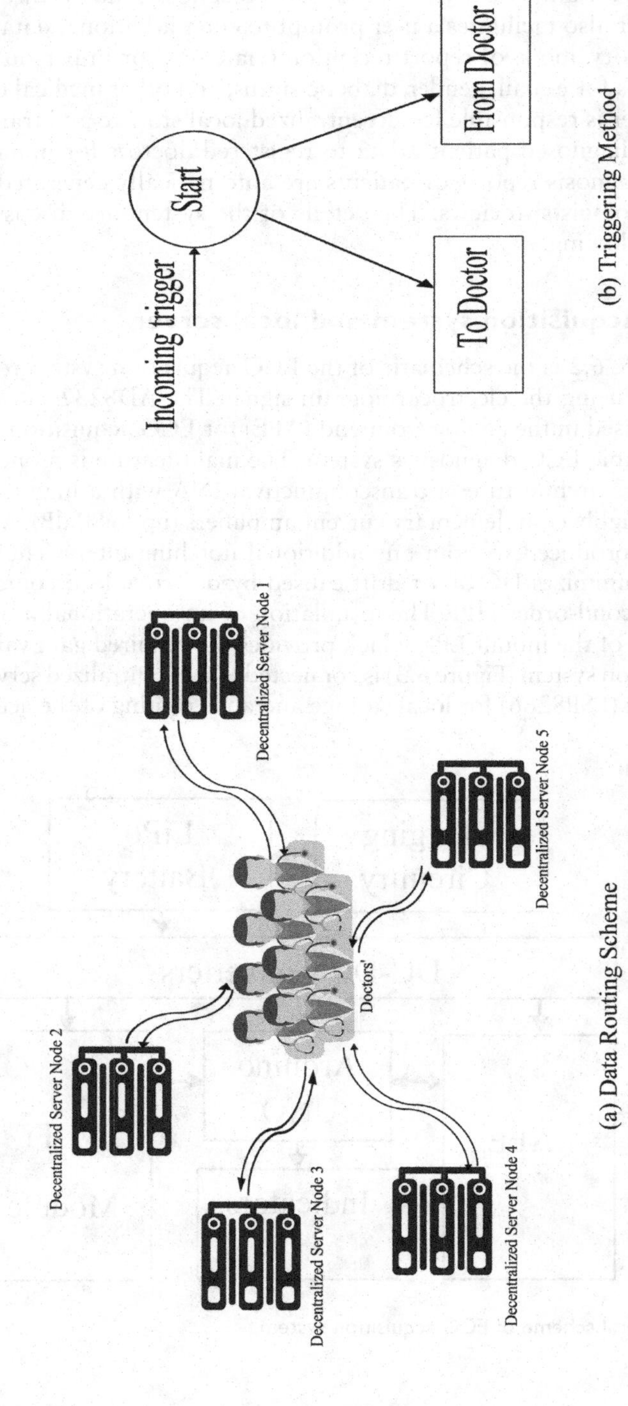

Figure 6.3 Structural scheme of ECG data routing from decentralized nodes to doctors.

ECG data for immediate assessment. A decentralized server is implemented using a Raspberry Pi 3B+ module installed with the Raspbian operating system. The local server is set up with the Apache server, along with the MySQL database management system. The Apache server is used for replicating the hosting service, whereas MySQL manages the registered doctors' information and patients' relational data extracted from different tags.

The server is configured to update the patient/doctor information when new entries are registered. The entry of new patient data triggers the integrated proposed algorithm for immediate assessment via forwarding the new data to the doctors. When the web server gets an HTTP POST request from a specific node, it first performs data frame authentication. It assigns a unique token to that node before copying the data supplied by the node to the server database. Meanwhile, the entry of a new doctor triggers the administrator verification. After successful validation, new doctors are added to the existing doctors' list. Details of the decentralized routing algorithm are discussed in Sections 6.2.2 and 6.2.3.

6.2.2 Decentralized data routing

Decentralized routing of undiagnosed patients' data is divided into two contexts, as shown in Figure 6.3. The route computing algorithm is triggered via two main activities: a new patient entry or a doctor response. When any trigger is activated, the routing algorithm gets activated to perform the appropriate routing routine.

As shown in Figure 6.3, when a new patient's data triggers the routing algorithm, the algorithm selects three doctors from the lists of registered doctors for quick diagnosis. The selection of doctors is based on some performance parameters, such as cooling period, responsiveness, and efficiency scores. The cooling period defines the duration of the last transaction from the doctor. The responsiveness score determines the time taken by the doctor to respond. The efficiency score defines the number of successful transactions. Meanwhile, when the doctors' response triggers the algorithm, tagged critiques from the data are extracted by the interface. A detailed patient report is constructed and forwarded to the appropriate patient.

6.2.3 Routing algorithm at node

Shown in Figure 6.4 (a) and (b) are the flowcharts representing the routing algorithm. Two conditions trigger the routing algorithm in the decentralized nodes; firstly, when new patient data is entered and, secondly, when critiques from a doctor are received. In the first case, when new patients' data is received in the decentralized node, the proposed algorithm tries to forward the data to specialists for immediate assessment via the To Doctor pathway.

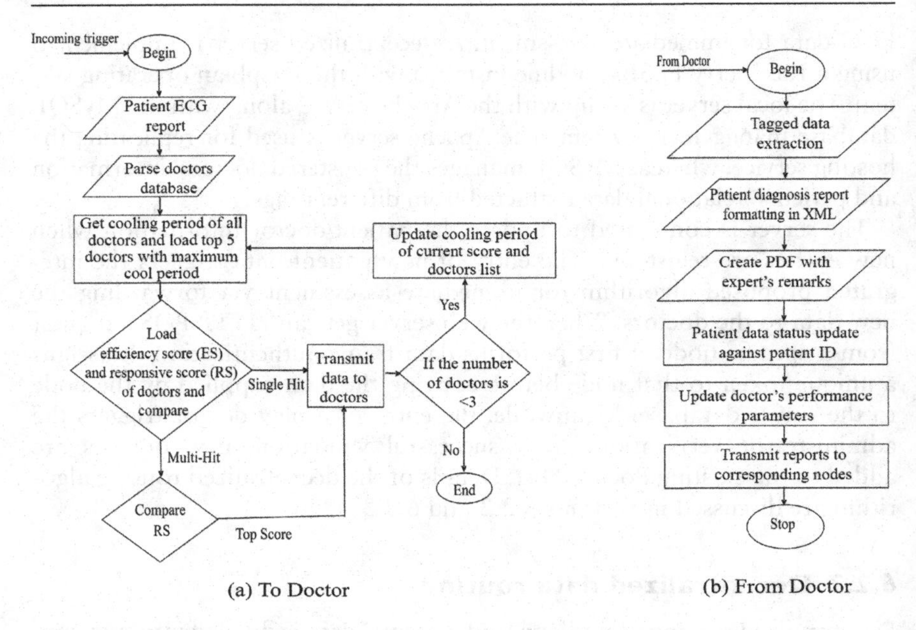

Figure 6.4 **Flowchart of the routing algorithm.**

In the To Doctor pathway, the algorithm accesses the doctors' database to forward the undiagnosed patient's data. The algorithm runs through the entire database to get the maximum value of the cooling period and sort the data with the highest cooling period first and the lowest cooling period last. The cooling period refers to the duration between the last completed transaction time and the current time. After sorting the doctors' database with the maximum cooling period, the algorithm searches for the doctor with the maximum efficiency score. If a single instance of the score is hit, then the ECG data is immediately forwarded to the doctor. In case of multiple hits with the same efficiency score, the algorithm searches for the minimum value of responsive scores, and the algorithm iterates to forward the data to a total of three doctors for immediate assessment.

Meanwhile, when the algorithm is triggered due to the incoming signal from the doctor, the flowchart representation of the algorithm in Figure 6.4 (b) is executed. After receiving the data from a doctor, the algorithm searches for the unique ID of the patient for a status check. If the status check shows the current status as Diagnosed, incoming data from a doctor is rejected, and the process stops; whereas an undiagnosed status of the patient will load the unique patients' database for update. The algorithm extracts the tagged data from the doctors' critique, and the data in the corresponding patient's database is updated with the assessment. The appropriate patients' database is identified from the unique ID tagged to the ECG data during the acquisition. Separate data in XML for the patient corresponding to a unique ID is also created, which is further used in patients' diagnostic report generation in PDF. After the report is generated, the patient's status is updated along with the new scores for the cooling period, responsiveness, and efficiency. The generated PDF report is forwarded to the patient if the patient's e-mail ID is provided for report delivery. All the doctor's assessments are stored in raw format, and a custom report for the three doctors can also be generated on demand for diagnostics comparison.

6.3 RESULTS AND DISCUSSION

Figure 6.5(a) depicts the onsite ECG data collection system. The procedure for obtaining an ECG is simple, and the data obtained from the patient has been pre-processed within the microcontroller to eliminate low-frequency, high-frequency, and powerline disturbances. Low-frequency noises from respiration contribute to the signal's baseline wander. Still, muscular noises and internal noises from the electronics component contribute to the signal's high-frequency noises, which must be eliminated before analysing ECG data. This clean ECG data is subsequently sent to the Wi-Fi module, which sends it to the local storage server. Figure 6.5(b) depicts the hardware setup of the recommended server. The numerous forms for report production from local server data are also shown in Figure 6.6, which allows the administrator to record extra information and tag the data with fields such as the patient's name, age, contact information, and other medical records.

As demonstrated in Figure 6.6(a), the administrative panel provides a variety of visual metrics, including the entire team of specialists, the total number of active doctors at any particular moment, and the total number of patients. When a new doctor registers in the system, the administrator is required to go to the incoming request section and approve the new doctor's requests to ensure their legality, as shown in Figure 6.6(b). As

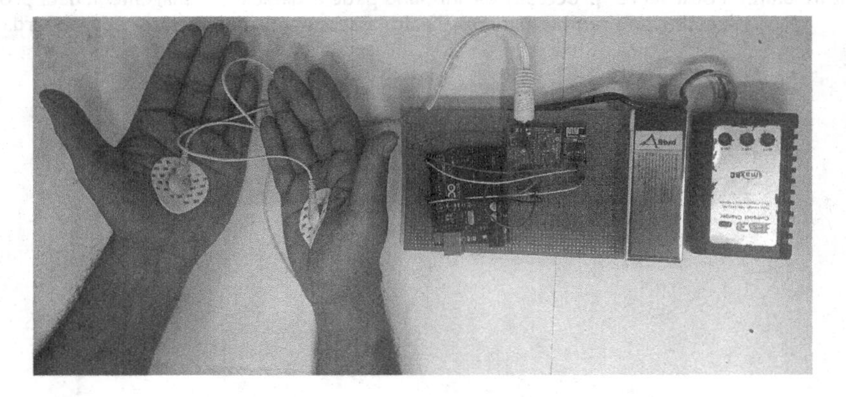

Figure 6.5(a) Data acquisition system.

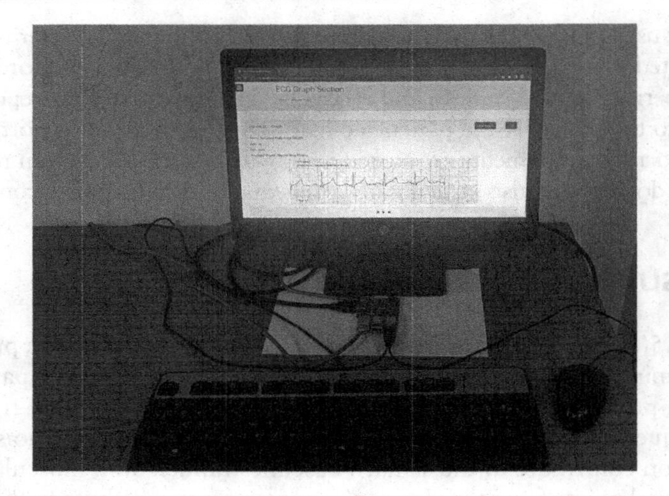

Figure 6.5(b) Hardware (server) setup of the proposed system.

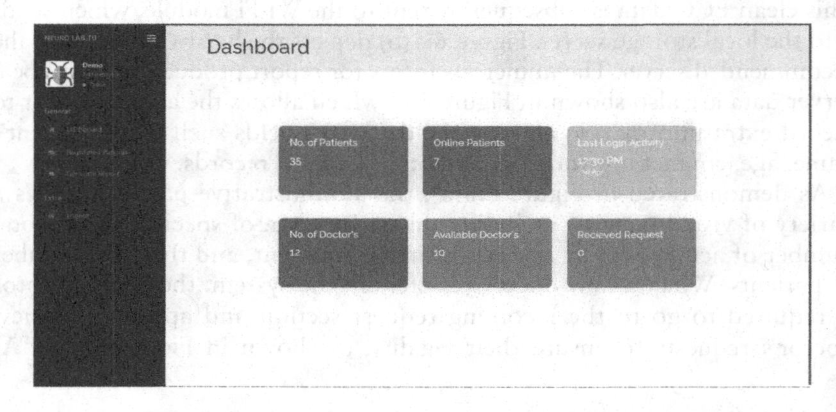

Figure 6.6(a) Local server processes for inbound patient database management, data processing, and remote patients' health assessment: administrators dashboard.

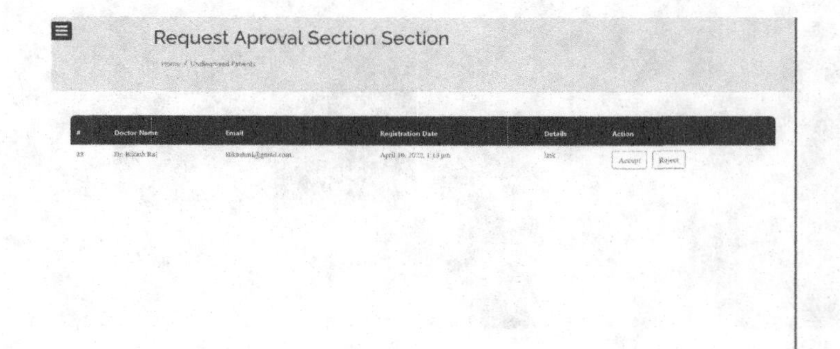

Figure 6.6(b) Admin approval request.

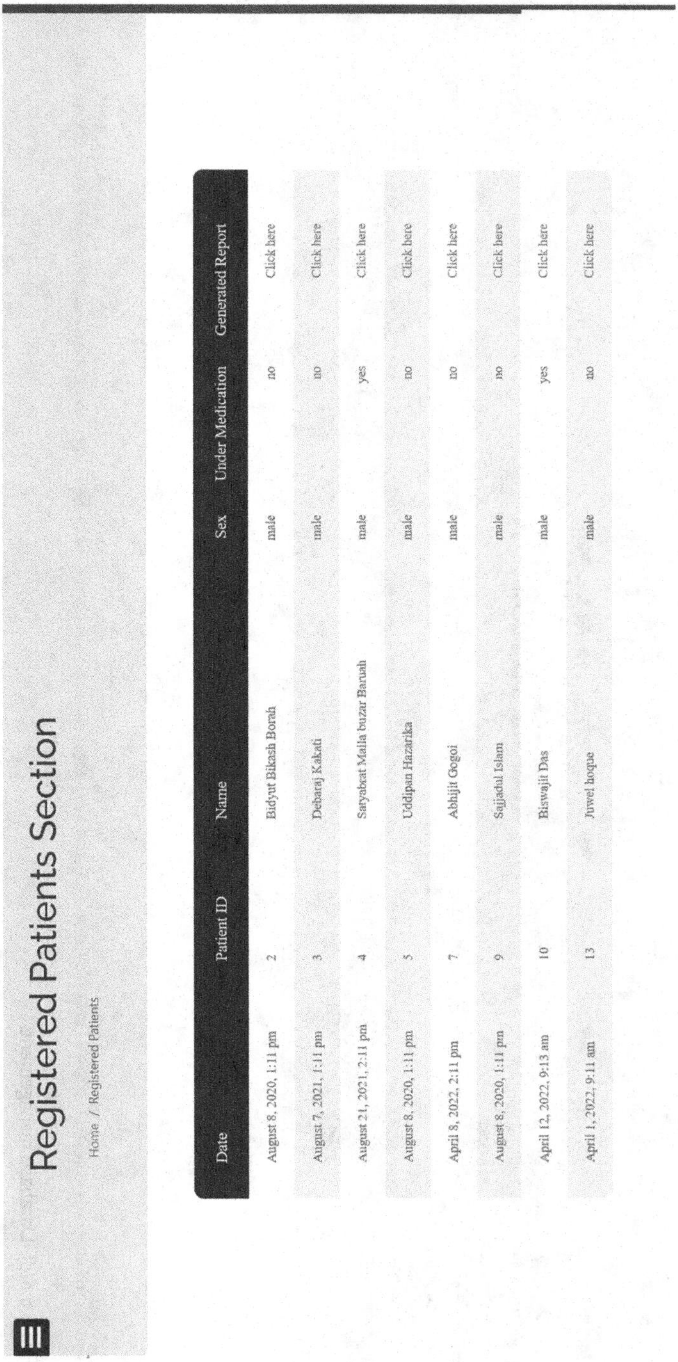

Registered Patients Section

Home / Registered Patients

Date	Patient ID	Name	Sex	Under Medication	Generated Report
August 8, 2020, 1:11 pm	2	Bidyut Bikash Borah	male	no	Click here
August 7, 2021, 1:11 pm	3	Debaraj Kakati	male	no	Click here
August 21, 2021, 2:11 pm	4	Satyabrat Maila buzar Baruah	male	yes	Click here
August 8, 2020, 1:11 pm	5	Uddipan Hazarika	male	no	Click here
April 8, 2022, 2:11 pm	7	Abhijit Gogoi	male	no	Click here
August 8, 2020, 1:11 pm	9	Sajjadul Islam	male	no	Click here
April 12, 2022, 9:13 am	10	Biswajit Das	male	yes	Click here
April 1, 2022, 9:11 am	13	Juwel hoque	male	no	Click here

Figure 6.6(c) Registered patients.

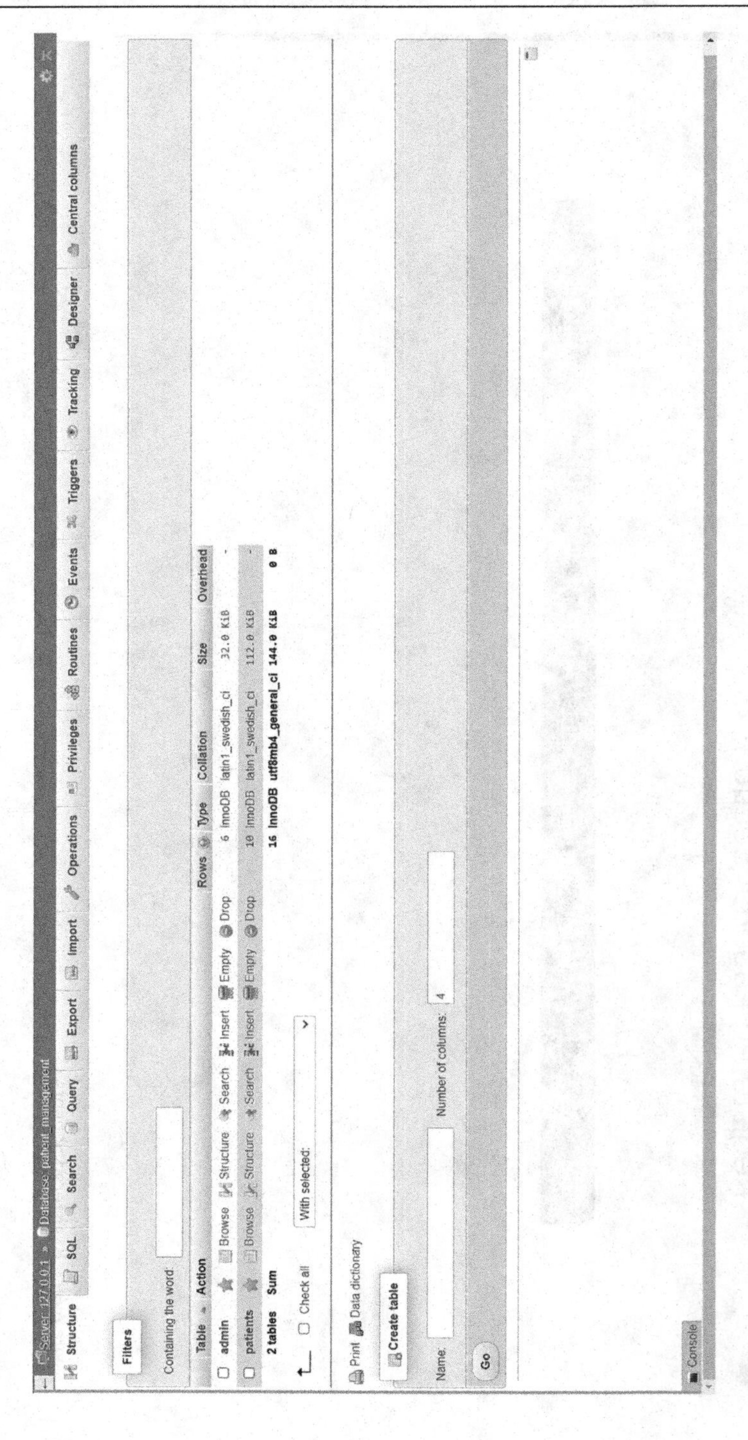

Figure 6.6(d) Database management.

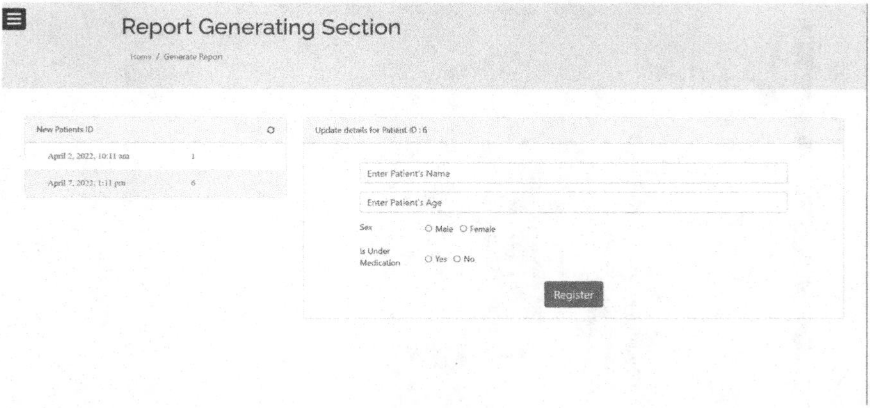

Figure 6.6(e) Patient information.

indicated in Section 6.2.1, when a new patient's data is received at the local server, the system displays a screen similar to that shown in Figure 6.6(e) for adding supplemental patient information. Two separate tables for patients and physicians are set up in the system repository. The data collected from the ECG collection unit will be recorded into the patient database using the acquisition unit's unique patient identifier, as depicted in Figure 6.6(f). As indicated in Figure 6.6(c), the administrator will be able to monitor a table, including all patient information. The system will automatically be able to route undiagnosed patients' ECG data to physicians as described in Section 6.2.3.

As displayed in Figure 6.6(h), the doctor's panel provides a variety of visible metrics, including the total number of assigned patients and the number of undiagnosed and diagnosed patients. Physicians may adjust their availability by clicking on the set availability option displayed in Figure 6.6(i). Finally, after the server receives the physicians' reports, it will automatically extract the doctors' remarks and provide a final diagnostic report, as seen in Figure 6.6(j).

6.4 CONCLUSION

The proposed setup comprises onsite medical ECG data acquisition equipment integrated with a local storage server that can store patients' ECG data and any additional information needed for crucial patient assessment. The device, connected to the internet, can relay data to experts in real time, allowing for a quick evaluation of the patients' severity. By connecting the local server through the internet, the present approach emphasizes local

Server: 127.0.0.1 » Database: patient_management » Table: patients

Browse | Structure | SQL | Search | Insert | Export | Import | Privileges | Operations | Tracking | Triggers

Profiling [Edit inline] [Edit] [Explain SQL] [Create PHP code] [Refresh]

Show all | Number of rows: 25 | Filter rows: Search this table | Sort by key: None

+ Options

	id	ecg	res	datetime	name	under_med	age	sex	is_generated	report	diagnoseBy
Edit Copy Delete	1	12,13,18,27,38,44,44,43,46,52,57,57,53,49,48,49,50...	101	April 2, 2022, 10:11 am	John Doe	no	32	Male	no	not_assign	13
Edit Copy Delete	2	13,18,38,44,44,43,46,52,57,57,53,49,48,49,50,51,54...	001	August 8, 2020, 1:11 pm	Bidyut Bikash Borah	no	23	male	yes	assign	3
Edit Copy Delete	3	13,18,38,44,44,43,46,52,57,57,53,49,48,49,50,51,54...	001	August 7, 2021, 1:11 pm	Debaraj Kakati	no	24	male	yes	assign	12
Edit Copy Delete	4	13,18,38,44,44,43,46,52,57,57,53,49,48,49,50,51,54...	001	August 21, 2021, 2:11 pm	Satyabrat Malla buzar Baruah	yes	35	male	yes	assign	5
Edit Copy Delete	5	13,18,38,44,44,43,46,52,57,57,53,49,48,49,50,51,54...	001	August 8, 2020, 1:11 pm	Uddipan Hazarika	no	23	male	yes	assign	9
Edit Copy Delete	6	12,13,18,27,38,44,44,43,46,52,57,57,53,49,48,49,50...	101	April 7, 2022, 1:11 pm	Mukesh Kumar	no	35	male	no	not_assign	35
Edit Copy Delete	7	13,18,38,44,44,43,46,52,57,57,53,49,48,49,50,51,54...	001	April 8, 2022, 2:11 pm	Abhijit Gogoi	no	25	male	yes	assign	28
Edit Copy Delete	9	13,18,38,44,44,43,46,52,57,57,53,49,48,49,50,51,54...	001	August 8, 2020, 1:11 pm	Sajjadul Islam	no	28	male	yes	assign	15
Edit Copy Delete	10	13,18,38,44,44,43,46,52,57,57,53,49,48,49,50,51,54...	001	April 12, 2022, 9:13 am	Biswajit Das	yes	30	male	yes	assign	23
Edit Copy Delete	13	13,18,38,44,44,43,46,52,57,57,53,49,48,49,50,51,54...	001	April 1, 2022, 9:11 am	Juwel hoque	no	29	male	yes	assign	6

Check all *With selected:* Edit Copy Delete Export

Show all | Number of rows: 25 | Filter rows: Search this table | Sort by key: None

Query results operations

Print | Copy to clipboard | Export | Display chart | Create view

Bookmark this SQL query

Console

Figure 6.6(f) Patients database structure.

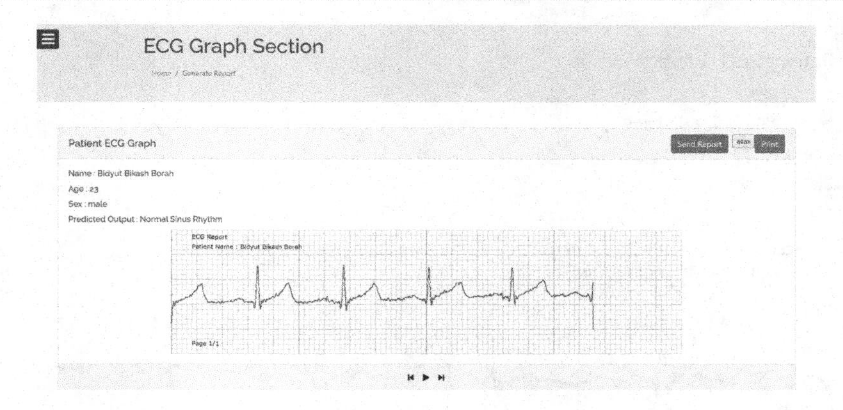

Figure 6.6(g) Saving patient report.

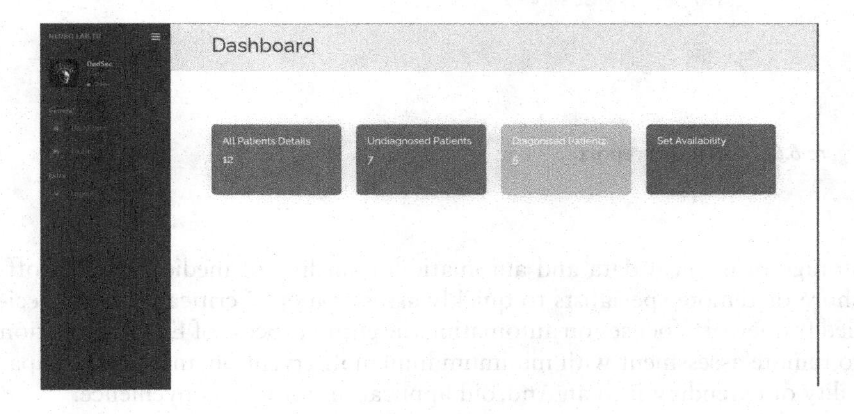

Figure 6.6(h) Doctors dashboard.

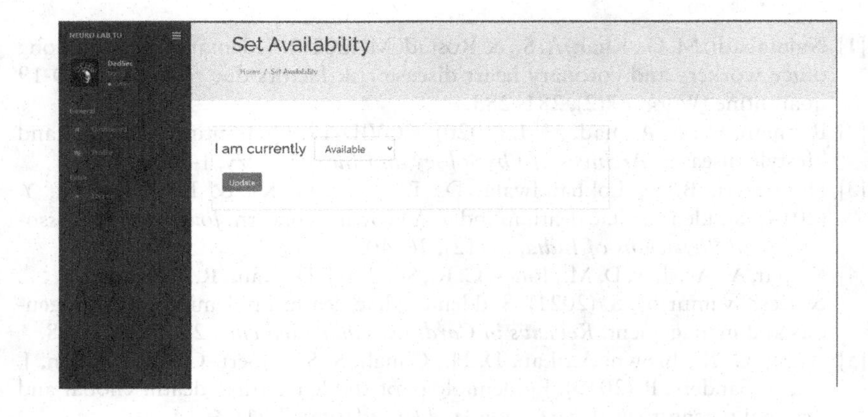

Figure 6.6(i) Doctors availability.

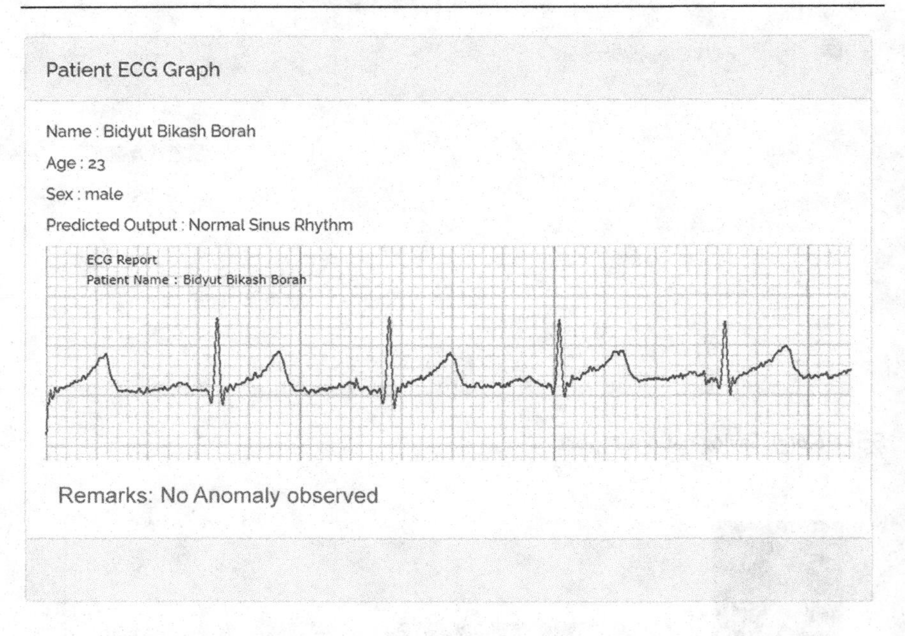

Figure 6.6(j) Diagnosis report.

storage of medical data and automatic forwarding of medical data to offshore or remote specialists to quickly assess patients' criticality. The specified framework focuses on automating the entire process of ECG acquisition to remote assessment with minimum human intervention and has the capability of extending it to an Android application for user convenience.

REFERENCES

[1] Najafabadi, M. G., Khah, A. S., & Rostad, M. (2020). Sedentary lifestyle among office workers and coronary heart disease risk factors due to the COVID-19 quarantine. *Work*, *67*(2), 281–283.

[2] Rehman, H., & Ahmad, M. I. (2020). COVID-19: Quarantine, isolation, and lifestyle diseases. *Archives of Physiology and Biochemistry*, 1–5.

[3] Honnekeri, B. S., Lokhandwala, D., Panicker, G. K., & Lokhandwala, Y. (2014). Sudden cardiac death in India: A growing concern. *Journal of the Association of Physicians of India*, *62*(12), 36–40.

[4] Kumar, A., Avishay, D. M., Jones, C. R., Shaikh, J. D., Kaur, R., Aljadah, M., . . . & Keshavamurthy, S. (2021). Sudden cardiac death: Epidemiology, pathogenesis and management. *Reviews in Cardiovascular Medicine*, *22*(1), 147–158.

[5] Wong, C. X., Brown, A., Lau, D. H., Chugh, S. S., Albert, C. M., Kalman, J. M., & Sanders, P. (2019). Epidemiology of sudden cardiac death: Global and regional perspectives. *Heart, Lung and Circulation*, *28*(1), 6–14.

[6] Savarese, G., & Lund, L. H. (2017). Global public health burden of heart failure. *Cardiac Failure Review*, *3*(1), 7.
[7] Shimokawa, H., Miura, M., Nochioka, K., & Sakata, Y. (2015). Heart failure as a general pandemic in Asia. *European Journal of Heart Failure*, *17*(9), 884–892.
[8] Velavan, P., Khan, N. K., Goode, K., Rigby, A. S., Loh, P. H., Komajda, M., . . . & Cleland, J. G. (2010). Predictors of short term mortality in heart failure—insights from the Euro Heart Failure survey. *International Journal of Cardiology*, *138*(1), 63–69.
[9] Beaton, A., Aliku, T., Okello, E., Lubega, S., McCarter, R., Lwabi, P., & Sable, C. (2014). The utility of handheld echocardiography for early diagnosis of rheumatic heart disease. *Journal of the American Society of Echocardiography*, *27*(1), 42–49.
[10] Prakash, S., & Venkatesh, V. (2013). Real time monitoring of ECG signal using PIC and web server. *International Journal of Engineering and Technology (IJET)*, *5*(2), 1047–1053.
[11] Xia, H., Asif, I., & Zhao, X. (2013). Cloud-ECG for real time ECG monitoring and analysis. *Computer Methods and Programs in Biomedicine*, *110*(3), 253–259.
[12] Harrison, J. P., & Lee, A. (2006). The role of e-health in the changing health care environment. *Nursing Economics*, *24*(6), 283.
[13] Raj, C., Jain, C., & Arif, W. (2017, March). HEMAN: Health monitoring and nous: An IoT based e-health care system for remote telemedicine. In *2017 International Conference on Wireless Communications, Signal Processing and Networking (WiSPNET)* (pp. 2115–2119). IEEE.
[14] Saikia, A., Sill, S., Baruah, S. M., Roy, S., & Das, B. R. (2020, July). Android interface for handling ECG data from GE MAC 600 ECG system. In *2020 International Conference on Computational Performance Evaluation (ComPE)* (pp. 771–776). IEEE.
[15] Maqbool, S., Iqbal, M. W., Naqvi, M. R., Arif, K. S., Ahmed, M., & Arif, M. (2020, November). IoT based remote patient monitoring system. In *2020 International Conference on Decision Aid Sciences and Application (DASA)* (pp. 1255–1260). IEEE.
[16] Tyagi, A. K., Gupta, M., Aswathy, S. U., & Ved, C. (2021). Healthcare solutions for the next generation: A useful explanation from the user's perspective. In *Recent Trends in Blockchain for Information Systems Security and Privacy* (pp. 175–217). CRC Press.
[17] Yew, H. T., Ng, M. F., Ping, S. Z., Chung, S. K., Chekima, A., & Dargham, J. A. (2020, February). IoT based real-time remote patient monitoring system. In *2020 16th IEEE International Colloquium on Signal Processing & Its Applications (CSPA)* (pp. 176–179). IEEE.

Chapter 7

Evaluation of deep image embedders for healthcare informatics improvement using visualized performance metrics

T. O. Olaleye, A. O. Okewale, I. Solanke, O. V. Alomaja, O. F. Adebayo, and S. M. Akintunde

CONTENTS

7.1 Introduction: Application of data science for healthcare informatics 115
7.2 Survey of related literature 117
7.3 Research methodology 118
7.3.1 Data acquisition 119
7.3.2 Image embedding through transfer learning 119
7.3.3 Machine learning implementation 122
7.4 Result and discussion 124
7.5 Conclusion and recommendation 132
Acknowledgement 133
References 133

7.1 INTRODUCTION: APPLICATION OF DATA SCIENCE FOR HEALTHCARE INFORMATICS

The application of machine learning for health informatics entails computational efforts before and after the data pre-processing phases of proposed frameworks. In a machine learning–based categorization of image signals, for a supervised learning approach, for instance, the extraction of feature vectors from the image input signals for analytics purposes remains germane and requires computational efforts. The extraction, as a use case of signal processing, is widely deployed in the research industry for a wide range of intents, including human biometrics [1], medical imaging [2], plant biometrics [3], and basic image differentiation [4]. The aforementioned use cases have returned state-of-the-art results through varying methodological frameworks and technicalities. In the prevalent era of pandemics, image data

DOI: 10.1201/9781003305347-7

comes in handy for research purposes in a concerted effort towards flattening the curve [5]. What's more, the Omicron variant of the Covid-19 is indeed worrisome; hence, there is a palpable fear of another global socioeconomic lockdown. The rapid upsurge of positive cases since the advent of the variant resulted in severe impacts on the mental and physical health of citizens around the world ever since the advent of the pandemic that dates back to December 2019 in Wuhan, China. Daily, several thousands of new cases are being reported, collated by the World Health Organization (WHO), and as of 3rd February 2022, a total number of 383,509,779 infected cases has been reported, alongside 5,693,824 recorded casualties [6]. The global health emergency status of the virus, therefore, requires a concerted effort by stakeholders towards taming the tide. Consequently, widespread recommendation of X-ray signals for Covid-19 analysis has been advocated lately, owing to the availability of X-ray machines in radiology departments of most public hospitals [7]. Widespread agitation for CT scan images for the diagnosis is likewise commonplace in literature, being reputed as a better choice for soft tissue–based ailments [7], and since X-ray machines cannot distinguish soft tissues reliably [7], CT scan signals are recommended as better suited for Covid-19 diagnosis. Image scanning tools notwithstanding, image feature extraction techniques are of paramount importance, as mentioned earlier, to the performance metrics of any diagnostic model. General image features including shape, texture, and color are less necessary for the Covid-19 use case of health informatics, unlike domain-specific features, and thereby require due consideration in pandemic diagnosis. Whereas several feature extraction techniques abound in medical imaging [8], the image-embedding feature vector–extracting approach is seldom deployed for Covid-19 medical imaging studies, which have demonstrated expertise in other use case domains. Image embedders read images and deploy deep learning models to calculate feature vectors for each image input. A data table of numeric image descriptors is consequently returned for further analysis. In most Covid-19 predictive studies, the deep learners are employed as learning models, rather than the vector-extraction tool approach of this chapter. Several pretrained deep embedders, including SqueezeNet [9], Inception v3 [10], VGG-19 [11], and VGG-16 [12] are either trained on ImageNet, Kaggle's Painter by Numbers set, Authors, and other prominent training sets. In this chapter, the expertise of the embedders is activated and evaluated on the input image set through a knowledge transfer mechanism to extract the numeric feature vectors. The performance of the embedders for feature extraction is further buoyed by the robustness of their varying training sets.

Consequent to the foregoing performance evaluation of the deep image embedders on medical imaging for health informatics, improvement is the focus of this chapter. CT scan and X-ray image signals on positive and negative cases of coronavirus are deployed as training sets and evaluated on five pretrained embedders for numeric feature extraction through knowledge transfer methodology to subsequently train three machine learning algorithms for instant Covid-19 diagnosis. As much as we know, this work is unique for the deployment of both X-ray and CT scan signals purposely for the evaluation of prominent pretrained deep embedders to evaluate their performances based on Covid-19 diagnosis. Statistical and visualized parameters are employed for the evaluation of the models, and the rest of the study is presented thus: Section 7.2 reviews similar studies and models, while Section 7.3 discusses the research methods. The experimental result is discussed in Section 7.4, while the deductions from the study are presented in Section 7.5.

7.2 SURVEY OF RELATED LITERATURE

There has been a surge of research efforts proposing different predictive models for medical imaging, thereby improving efforts towards health informatics, especially on the Covid-19 debacle. Several such seminal efforts are addressed in this section. In [7], a bi-stage feature selection approach is implemented by Sen et al. to identify the most significant feature attributes extracted from CT scan images, comprising positive and negative cases. The study trained the secure vector machine classifier with two public sets, returning a 98.39% and 90.0% accuracy rates, respectively. In [13], Zhang et al. deployed the ResNet model to diagnose coronavirus cases on a chest X-ray set of 1,078 images, returning a 96% accuracy rate. Wang L. et al. deployed CNN for its detection system, with a 92.6% accuracy in [14]. Their training set comprises X-ray signals of pneumonia, Covid-19, and negative cases. In a similar study to our approach, Narin et al., in [15], deployed popular deep learning models, including ResNet-50 and Inception-ResNetV2 on X-ray image signals. A class decomposition technique is used in the work of Abbas et al. in [2] to discover irregularities in chest X-ray signals. The scheme could recognize Covid-19 infection with 95.12% accuracy.

Feature vectors from CXR images were extracted through the instrumentality of deep learning by [16]. A secure vector machine is trained on the extracted numeric feature attributes for subsequent Covid-19 diagnosis. In the work of [17], a distant domain transfer learning methodology is presented on medical imaging to predict Covid-19 infection. Unlabelled training sets of X-ray images were employed with a classification accuracy of 96%, 13% higher than the non-transfer counterpart. CT scan was the training set for the work of [18], with a deep learner on the COVNet embedded training set, to identify Covid-19 and pneumonia. The study was appraised with the operating characteristic curve, area under the curve (AUC), and sensitivity. The study of Rehouma in [19] employed image signals for diagnosis through the instrumentality of transfer learning for deep learning. Models used include AlexNet, VGG, ResNet, GoogleNet, and Xception, with state-of-the-art experimental results. The study of [20] deployed eight models for the diagnosis of the infection through X-ray image signals using hyperparameter tuning. An accuracy of 98.33% is returned by ResNet-34. In [21], a deep Bayes-SqueezeNet model was designed. The model was aimed at predicting Covid-19 infection over the evaluation set, and the experiment returned state-of-the-art results. In [22], the study proposed a medical imaging solution, which is conceptualized as a web-based platform for the prediction of prostate cancer in MRI signals. The solution was executed by a deep learner algorithm with insignificant computational overheads. Another machine learning for medical imaging model was introduced by [23] using a convolutional neural network. The transfer learning approach returned good performance metrics, while [24] implemented a similar model with the functionality of Python programming language. The computational performance methodology is benchmarked against other related open-source frameworks, and the model is evaluated on a pediatric pneumonia test set. The work of [25] intends to demystify the high rate of sensitivity achieved by predictive models on Covid-19, deploy a heterogeneous training set under a close collaboration with a medical institution, and deploy a smart databased network methodology for state-of-the-art improvement. An accuracy of 97.72% was achieved from the experimental result. The study of [26] employed a transfer learning methodology through deep learning on publicly available X-ray image signals of 5,000 instances. Four networks were trained with sensitivity and specificity, with weighted averages of 98% and 90%, respectively. Prediction of the severity level of the virus and pneumonia on X-ray signals was achieved by fitting a deep neural network learner to determine the severity of the infection in the study [27]. Other instances of X-ray image signals were employed in the work of [28] for predicting the occurrence of coronavirus using a deep network. The assessment of the model with VGG-16 and AlexNet proves that supervised learning results for more than two classes appear more closely related and equal to three-class classifications.

7.3 RESEARCH METHODOLOGY

Similar to the study of Olaleye et al. in [3] that evaluates the performances of image filters, this chapter appraises the performance of image-embedding feature extractors, deployed through transfer learning, to ascertain their precision rate on medical imaging in the health informatics use case. It intends to determine the best learner algorithm fitted on the feature vectors extracted from image signals of positive and negative coronavirus cases and, likewise, the implication of the size of extracted feature attributes on performance. Recent literature has deployed image embedders for several purposes, including medical imaging, thereby employing either X-ray or CT scan image inputs. This study, however, combines both signals as a training set and implements the framework depicted in Figure 7.1 to actualize the aims of the study. The Orange data mining toolkit is employed in the implementation of the framework.

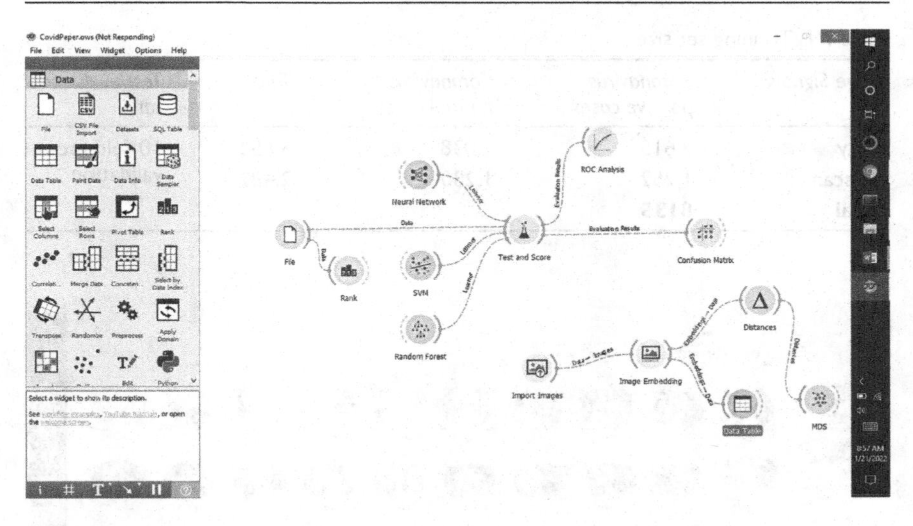

Figure 7.1 Framework of the study.

7.3.1 Data acquisition

The medical image signals are acquired as training evaluation sets. The signals are segmented sets from public repositories. The chest X-ray radiography set, obtained from [29], is spread across positive and negative instances, as described in Table 7.1. The CT scan version is acquired from [30] with an 8.8 usability rate covering 1,252 coronavirus positive and 1,230 negative instances. The final training set for this study returns 8,135 instances of both image signals, which will be subjected to feature extraction by the deep image embedders of SqueezeNet, Inception v3, VGG16, VG19, and Painters. Each embedder extracts different sizes of features from each of the 8,135 samples, which serves as the training set for this study. Table 7.2 shows the characteristics of the image embedders, with their training set sizes. The expected size of attributes from each of the embedders is likewise stated alongside other information. Screenshots of both signals are presented in Figure 7.2 and Figure 7.3, together with their corresponding extracted feature vectors by SqueezeNet.

7.3.2 Image embedding through transfer learning

Image embedding involves the passing of image signals through existing pre-trained deep networks to acquire the vector representations of the image input. The extracted vector representations are achieved through transfer learning either on a dedicated server or a local server. The aforementioned biomedical images, comprising X-ray and CT scan signals, are profiled by five deep convolutional image embedders with vectors of features for predictive analytics

Table 7.1 Training set size

Image Signals	*Coronavirus positive cases*	*Coronavirus normal cases*	*Total*	*Test/evaluation ratio*
X-ray	3,615	2,038	5,653	10-fold cross-validation
CT scan	1,252	1,230	2,482	
Total	**8135**			

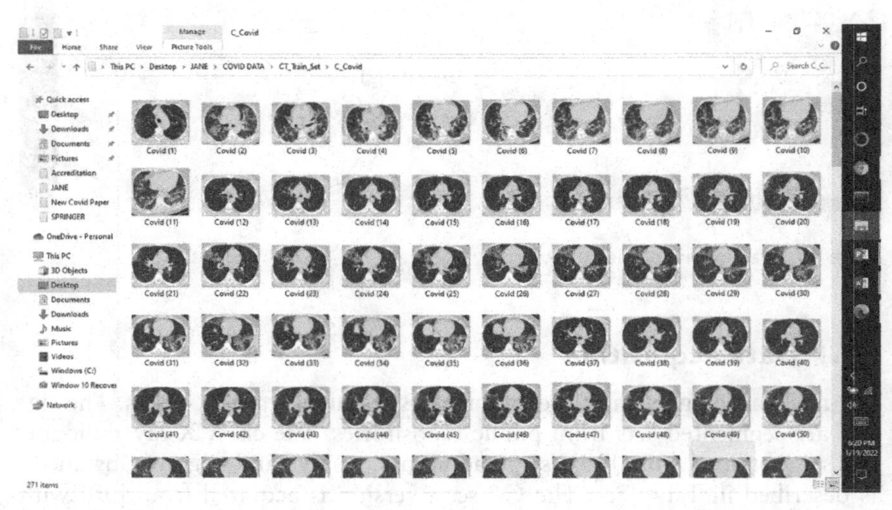

Figure 7.2 CT scan input signal with its corresponding feature vectors from image embedding.

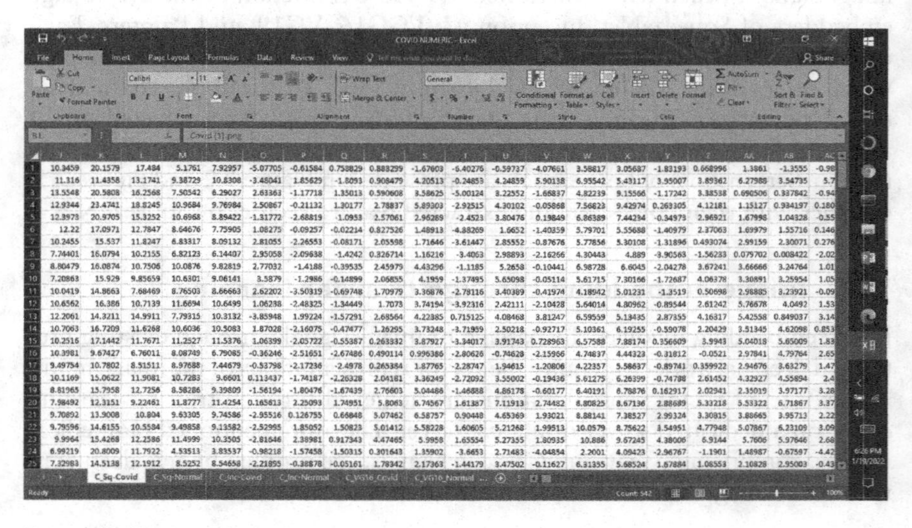

Figure 7.2 (Continued).

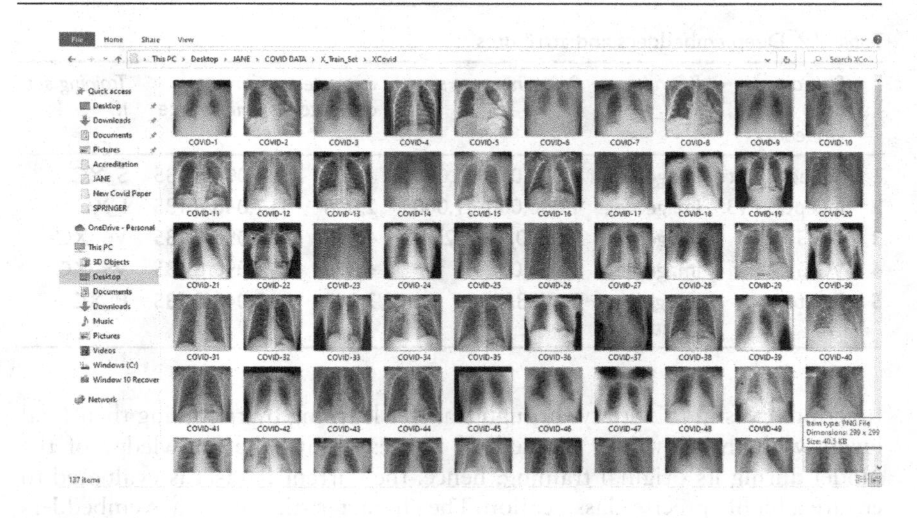

Figure 7.3 X-ray input signals with their corresponding feature vectors from image embedding.

	K	L	M	N	O	P	Q	R	S	T	U	V	W	X	Y	Z	AA	AB	AC	AD
1	0.744351	0	0	1.59561	1.51936	0	0	0.075524	0.554657	0	1.91681	0	0	6.63919	0	0	2.58837	0.845419	0	0.388
2	0	0	0	0.962494	0.758576	0	0	0.081413	0.657271	0	1.88872	0	0	3.73574	0.187111	0	0.917771	0.258515	0	0.415
3	2.05777	0	0	1.78469	1.79523	0	0	0.197368	0.63144	0	1.7846	0	0	4.98689	0	0	1.20218	0.80405	0	0.719
4	1.24438	0	0	1.46225	0	0	0	0.15493	0.560516	0	2.01144	0	0	5.64273	0	0	2.05681	1.03165	0	1.93
5	0.838925	0	0	2.02134	1.06296	0	0	0.126553	0.565377	0	2.12181	0	0	6.38632	0	0	2.6776	1.29126	0	0.509
6	0.595259	0	0	1.85207	1.46561	0	0	0.110158	0.586275	0	2.07079	0	0	5.76645	0	0	2.42504	0.877693	0	0.545
7	0.437093	0	0	1.98675	1.51894	0	0	0.088036	0.579274	0	2.25759	0	0	5.6596	0	0	2.66272	0.770686	0	0.528
8	0.353755	0	0	1.74027	1.76926	0	0	0.123032	0.615587	0	2.14632	0	0	4.6819	0	0	2.11131	0.620048	0	0.228
9	1.00089	0	0	1.82282	1.5911	0	0	0.06201	0.58169	0	2.00269	0	0	5.78452	0	0	2.52336	0.872726	0	0.727
10	0.953639	0	0	1.7246	1.55484	0	0	0.080482	0.585619	0	1.88871	0	0	5.77675	0	0	2.46382	0.926937	0	0.645
11	1.58407	0	0	1.7126	1.44614	0	0	0.05927	0.596765	0	1.74193	0	0	5.93366	0	0	2.11105	1.01772	0	0.590
12	2.12044	0	0	1.9172	1.80874	0	0	0.071788	0.559068	0	1.85942	0	0	6.53968	0	0	2.04385	1.11299	0	0.732
13	0	0	0	1.12415	0.257215	0	0	0.040806	0.631261	0	1.95945	0	0	3.50458	0.066851	0	1.22605	0.417845	0	0.538
14	1.16405	0	0	1.4803	1.22133	0	0	0.083423	0.59292	0	1.7358	0	0	5.53327	0	0	1.75392	1.02551	0	0.588
15	1.48272	0	0	1.68846	1.05984	0	0	0.07069	0.55865	0	1.74358	0	0	6.18085	0	0	2.13085	1.16317	0	1.13
16	1.52858	0	0	1.76614	1.79348	0	0	0.119968	0.609088	0	1.94388	0	0	4.8892	0	0	1.35147	0.939183	0	0.183
17	1.26794	0	0	1.75596	2.12272	0	0	0.195407	0.625689	0	1.93525	0	0	4.78531	0	0	1.2801	0.856041	0	0.260
18	1.13434	0	0	1.60293	2.34925	0	0	0.231739	0.646614	0	1.88686	0	0	4.72624	0	0	1.03045	0.70364	0	0.039
19	1.56613	0	0	1.55846	1.61906	0	0	0.172562	0.654643	0	1.90754	0	0	4.33758	0	0	1.25027	0.781033	0	0.086
20	1.34961	0	0	1.9488	0.910152	0	0	0.085188	0.575075	0	1.87832	0	0	5.30519	0	0	1.91973	1.07395	0	1.19
21	1.47703	0	0	2.06574	0.871896	0	0	0.079816	0.586819	0	1.96577	0	0	5.30616	0	0	2.10853	1.13238	0	1.16
22	1.1382	0	0	1.87259	1.36951	0	0	0.134611	0.613629	0	1.88471	0	0	4.74142	0	0	1.5681	0.910772	0	0.640
23	1.37133	0	0	1.82122	0.869804	0	0	0.117671	0.599793	0	1.9038	0	0	4.78865	0	0	1.83785	1.02617	0	1.00
24	0.30329	0	0	1.38669	0.710224	0	0	0.090255	0.62634	0	2.02288	0	0	3.74102	0	0	1.17799	0.556881	0	0.62
25	1.03585	0	0	1.82082	1.49795	0	0	0.171741	0.620077	0	1.98111	0	0	4.46712	0	0	1.55902	0.941706	0	0.548

C_Paint_Normal | C_Deep_Normal | C_Deep_Covid | C_Face_Covid | C_Face_Normal

Figure 7.3 (Continued)

for performance evaluation of the embedders for health informatics. The concept of transfer learning permits the training of the final layer of an existing deep model which then ensures little training time and, likewise, reduces data requirements and computational overheads. The five deep embedders employed for this study, as captured in Table 7.2, were originally trained on millions of training sets, with varying degrees of class sizes. Deploying their

Table 7.2 Deep embedders and attributes

	Pretrained deep embedder	*Pretraining set*	*Pretraining set size*	*Extracted attributes*	*Current training size*	*Training set ID*
1	SqueezeNet	ImageNet	1,000,000+/1,000	1,000	1,000/8,135	S_XC
2	Inception v3	ImageNet	1,000,000+/1,000	2,048	2,048/8,135	I_XC
3	VGG16	ImageNet	14,000,000/22,000	4,096	4,096/8,135	V6_XC
4	VGG19	ImageNet	1,000,000	4,096	4,096/8,135	V9_XC
5	Painters	Kaggle Painters	79,433/1,584	2,048	2,048/8,135	P_XC

functionalities in this study through transfer learning by retraining their final layers with the Covid-19 biomedical images retains the knowledge of the model during its original training; hence, the current dataset is evaluated to ensure a highly precise classification. The characteristics of the five embedders employed in this study are presented in the first four columns of Table 7.2. The basic architecture for all image embedders is premised on a deep learning methodology with input layers, hidden layers, and the subsequent output layers. The output layers of the five embedders return feature vector spaces of 300 attributes each on every input of a CT scan and X-ray image signal.

7.3.3 Machine learning implementation

This evaluation study is implemented by predictive analytics of the image inputs on three base learners, including secure vector machine (SVM), random forest (RF), and multilayer perceptron (MLP) deep learner. Each of the learners would be fitted on the S_XC, I_XC, V6_XC, V9_XC, and P_XC training sets, as depicted in the last column of Table 7.2, achieved through transfer learning. The machine learning methodology adopted in this study is implemented through 10-fold cross-validation, as described in Figure 7.4, on the Orange data mining toolkit. In the cross-validation, the embedder_attribute:8135 (attribute: instance) dataset is illogically allotted into 10 equivalent subsamples where one subsample serves as authentication. The three learner algorithms of SVM, RF, and MLP are treated the same way, and the cross-validation is repeated in 10 clock-folds. The results from each fold are subsequently computed to produce a distinct estimate for each of the learner algorithms. This approach guarantees all embedder_attribute:8135 pairs are used for the training and test set.

RF depicts an ensemble of decision trees with a bagging methodology. As observed in Figure 7.5, where a is the threshold, fv is the feature vector, and cs is the binary class (Covid-19 status), each node and its threshold are chosen to identify a minimal variety of possibilities in the children node. The diversity is measured by the Gini criterion which ensures the continuation of

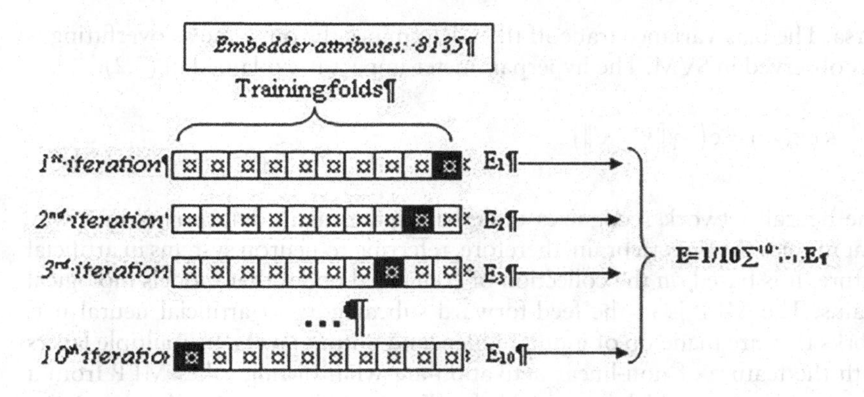

Figure 7.4 Concept of 10-fold cross-validation.

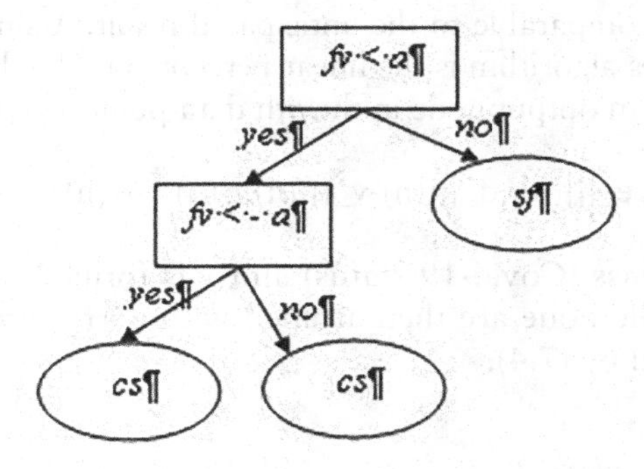

Figure 7.5 Random forest structure.

subdivisions until each node at the bottom of the tree is representative of one unique class. SVM is regarded as an extended version of the maximum margin classifier and is most applicable for overlapping classes and non-linear decision. The kernel function is a tuning parameter in an SVM model which removes the computational requirements for a higher dimensional vector space. A polynomial function kernel uses a degree 2 to identify non-linear data through transformation into higher dimensions, as in equation (7.1):

$$k(x, x') = (1 + x^{*}x')^{K} \tag{7.1}$$

While the radial basis kernel function, or the Gaussian, is capable of producing infinite dimensions to separate non-linear data, which is a function of a hyperparameter γ that must be scaled during data normalization. The bias becomes smaller when the hyperparameter is small, hence, a higher variance and vice versa. The bias-variance tradeoff then determines the outcome as overfitting is less observed in SVM. The hyperparameter impact is explained in (7.2):

$$k(x,x') = e\left(-\gamma\|x - x'\|\right) \tag{7.2}$$

The neural network recognizes underlying patterns in data points in a way that mimics the human brain, therefore, referring to neuron systems in artificial nature. It is based on the collection of connected nodes that models biological brains. The MLP is in the feed-forward subcategory of artificial neural networks that are made up of input, hidden, and output layers. Its multiple layers with the feature of non-linear activation are what distinguishes MLP from a linear perceptron, which is reputable for distinguishing and classifying non-linearly separable datasets. The learning takes place in the perceptron by altering linking weights immediately after each data point is treated which is a function of the error rate comparable to the anticipated result. A simplification of the least mean squares algorithm is the linear perceptron. The degree of error can be represented in an output node in the *n*th data point by *e(n)* in (7.3):

{\displaystyle e_{j}(n)=d_{j}(n)-y_{j}(n)}$e_j(n) = d_j(n) - y_j(n)$ (7.3)

Where d is the class (Covid-19 status) and y is formed by the perceptron. The weights of the node are then attuned with corrections by minimizing the error achieved by (7.4):

$$£(n) = \frac{i}{2}\sum_{j=0} e(n) \tag{7.4}$$

The change in each weight is computed in (7.5) using gradient descent:

$$\Delta wji(n) = -\eta\left(\frac{\eth\Sigma(n)}{\eth vj(n)}\right) yi(n) \tag{7.5}$$

and yi represents the result of the previous neuron, with η as the knowledge rate. The knowledge rate is designated to guarantee the weights eventually converge, without fluctuations, to a reaction. The computed derivative largely depends on local field vj.

7.4 RESULT AND DISCUSSION

This section discusses the experimental result of the evaluation study as implemented on the Orange data mining toolkit with Python libraries. The MLP, SVM, and RF algorithms were all evaluated (using 10-fold cross-validation) on the feature vector datasets described in the last column of

Table 7.2 which were extracted by SqueezeNet, Inception v3, VGG16, VGG19, and Painters, respectively. The performance metrics of the MLP, SVM, and RF models on the five datasets, S_XC, I_XC, VG6_XC, VG9_XC, and P_XC, are presented in Table 7.3, Table 7.4, Table 7.5, Table 7.6 and Table 7.7. Furthermore, the receiver operating characteristics (ROC) curve of each model, per dataset, is presented in Figure 7.6 for each of the

Table 7.3 Performance of models on SqueezeNet data attributes

Dataset/ training size	*Model*	*AUC*	*CA*	*F1*	*Precision*	*Recall*
S_XC [1,000/8,135]	RF	0.960	0.907	0.907	0.907	0.907
	SVM	0.999	0.992	0.992	0.992	0.992
	MLP	0.998	0.989	0.989	0.989	0.989

Table 7.4 Performance of models on inception v3 data attributes

Dataset/ training size	*Model*	*AUC*	*CA*	*F1*	*Precision*	*Recall*
I_XC [2,048/8,135]	RF	0.912	0.840	0.840	0.840	0.840
	SVM	0.992	0.958	0.958	0.958	0.958
	MLP	0.989	0.950	0.950	0.950	0.950

Table 7.5 Performance of models on VGG16 data attributes

Dataset/training size	*Model*	*AUC*	*CA*	*F1*	*Precision*	*Recall*
VG6_XC [4,096/8,135]	RF	0.975	0.921	0.921	0.921	0.921
	SVM	0.997	0.974	0.974	0.974	0.974
	MLP	0.997	0.997	0.997	0.997	0.997

Table 7.6 Performance of models on VGG19 data attributes

Dataset/ training size	*Model*	*AUC*	*CA*	*F1*	*Precision*	*Recall*
VG9_XC [4,096/8,135]	RF	0.969	0.903	0.903	0.903	0.903
	SVM	0.991	0.959	0.959	0.959	0.959
	MLP	0.991	0.968	0.968	0.968	0.968

Table 7.7 Performance of models on painters data attributes

Dataset/training size	*Model*	*AUC*	*CA*	*F1*	*Precision*	*Recall*
P_XC [2,048/8,135]	RF	0.999	0.988	0.988	0.988	0.988
	SVM	1.000	0.998	0.998	0.998	0.998
	MLP	1.000	0.998	0.998	0.998	0.998

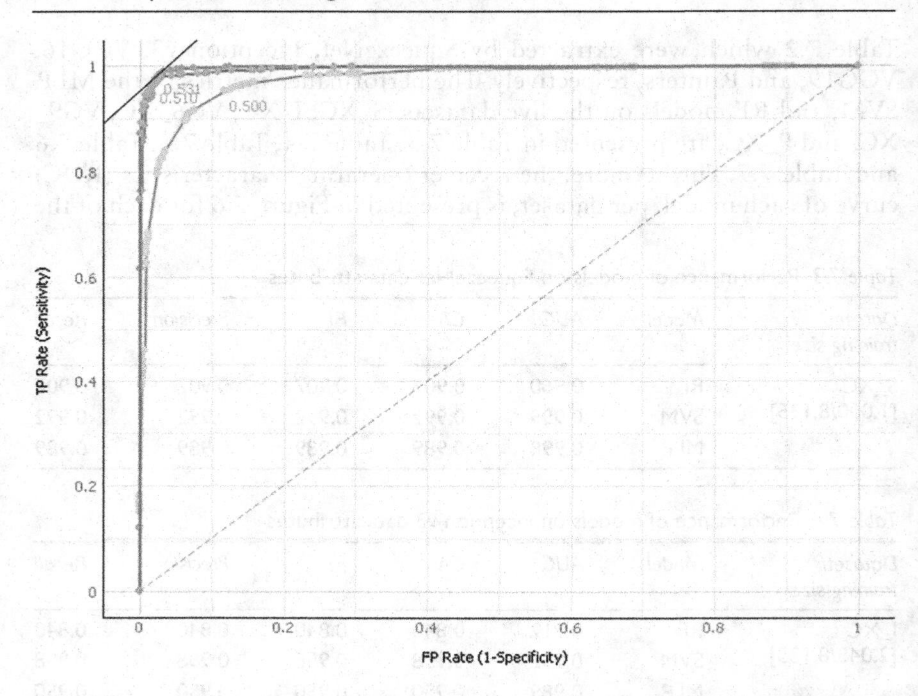

Figure 7.6(a) ROC curves of the three models on the five datasets: SqueezeNet ROC plot.

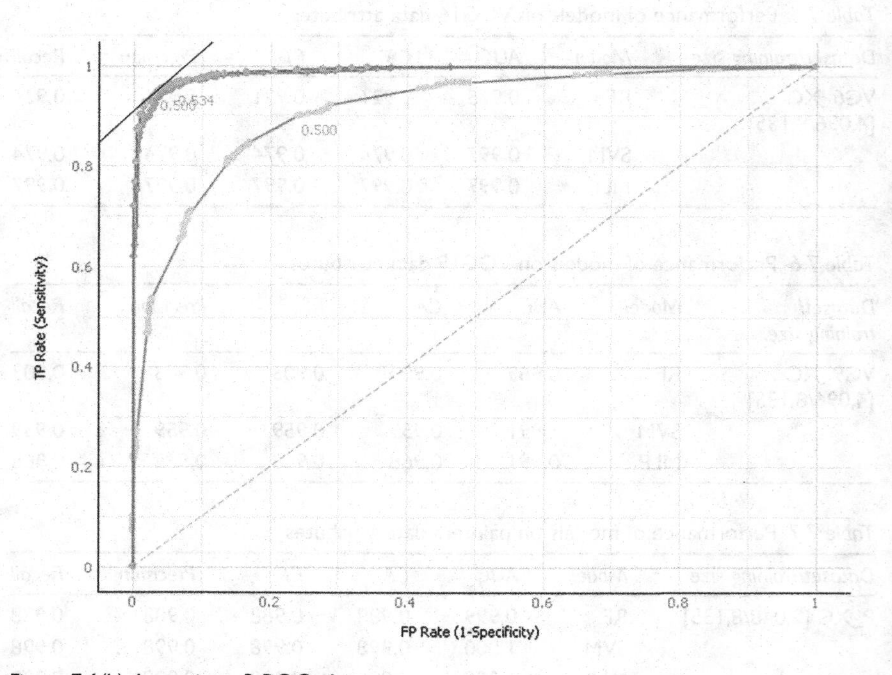

Figure 7.6(b) Inception v3 ROC plot.

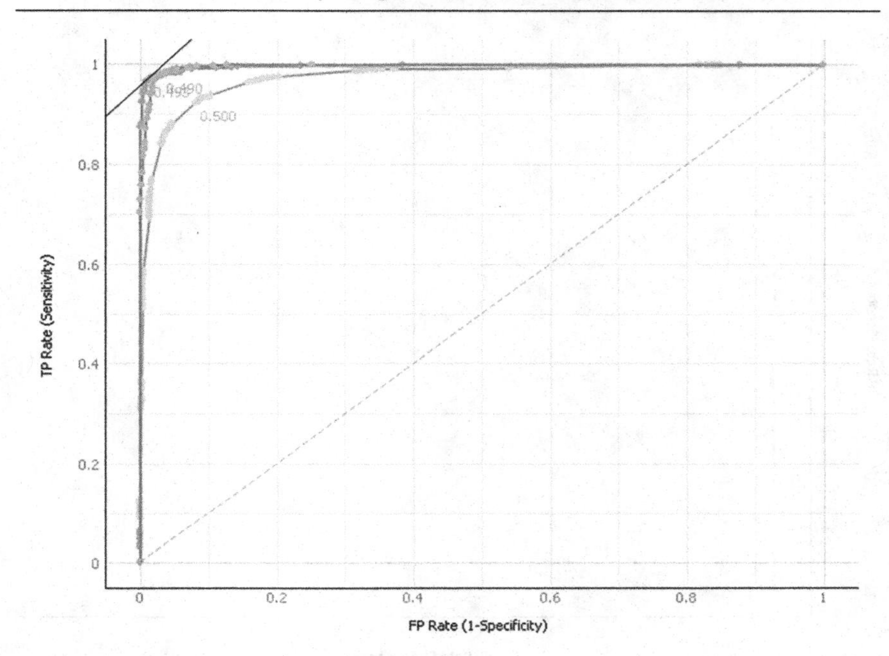

Figure 7.6(c) VGG16 ROC plot.

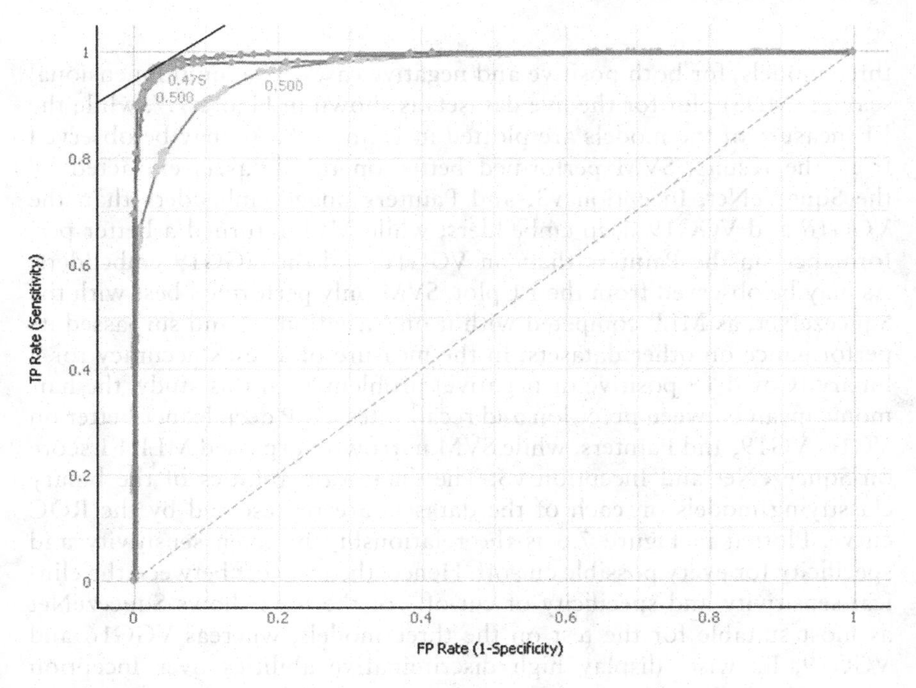

Figure 7.6(d) VGG19 ROC plot.

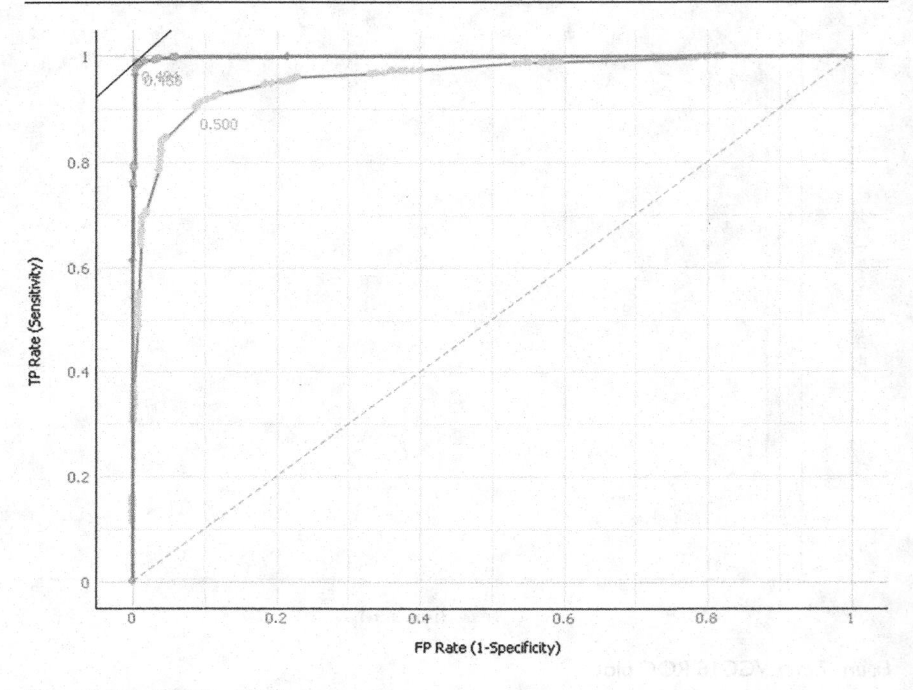

Figure 7.6(e) Painters ROC plot.

three models, for both positive and negative cases. The multidimensional scaling (MDS) plot for the five datasets is shown in Figure 7.7, while the F1 measure of the models are plotted in Figure 7.8. As may be observed from the results, SVM performed better on the datasets extracted by the SqueezeNet, Inception v3, and Painters image embedders than the VGG16 and VGG19 deep embedders; while MLP returned a better performance on the Painters than on VGG16 and the VGG19 embedders. As may be observed from the F1 plot, SVM only performed best with the SqueezeNet, as MLP competed with it on Inception v3 and surpassed its performance on other datasets. In the measure of a test's accuracy for a binary (Covid-19 positive or negative) problem as in this study, the harmonic mean between precision and recall rates MLP deep leaner better on VG16, VG19, and Painters, while SVM narrowly surpassed MLP F1 score on SqueezeNet and Inception v3. The diagnostic abilities of the binary classifying models on each of the datasets are represented by the ROC curve. Plotted in Figure 7.6 is the relationship between sensitivity and specificity for every possible cut-off. Hence, the tradeoff between the clinical sensitivity and specificity of cut-offs of the tests shows SqueezeNet as most suitable for the test on the three models, whereas VGG16 and VGG19, likewise, display high discriminative abilities over Inception

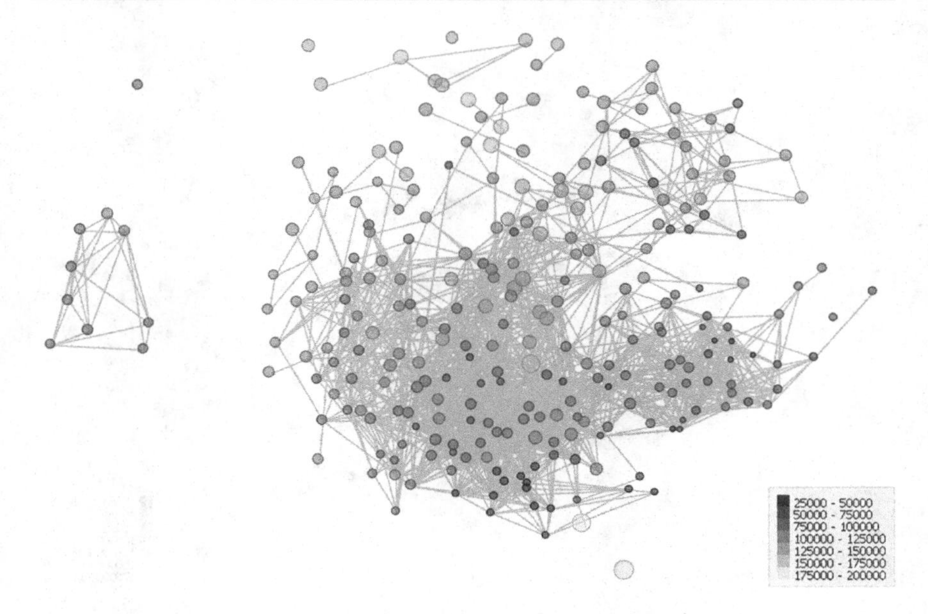

Figure 7.7(a) MDS plot of the five datasets: MDS plot for SqueezeNet.

Figure 7.7(b) MDS plot for Inception v3.

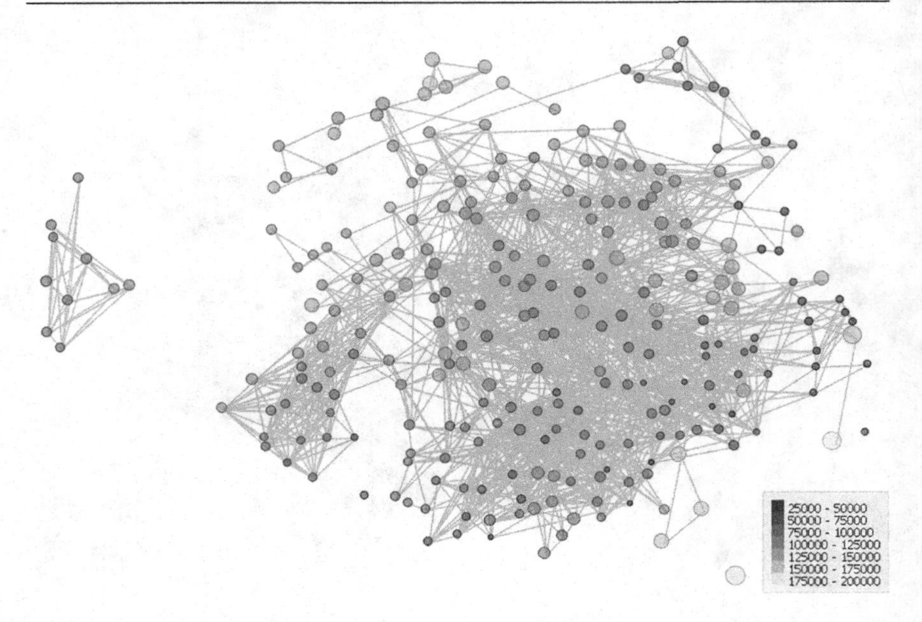

Figure 7.7(c) MDS plot for VGG16.

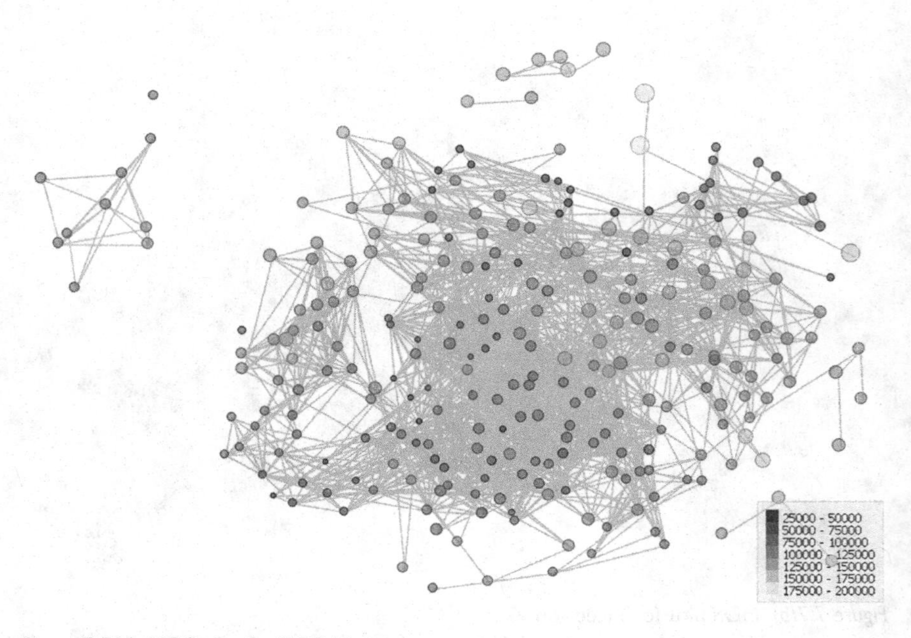

Figure 7.7(d) MDS plot for VGG19.

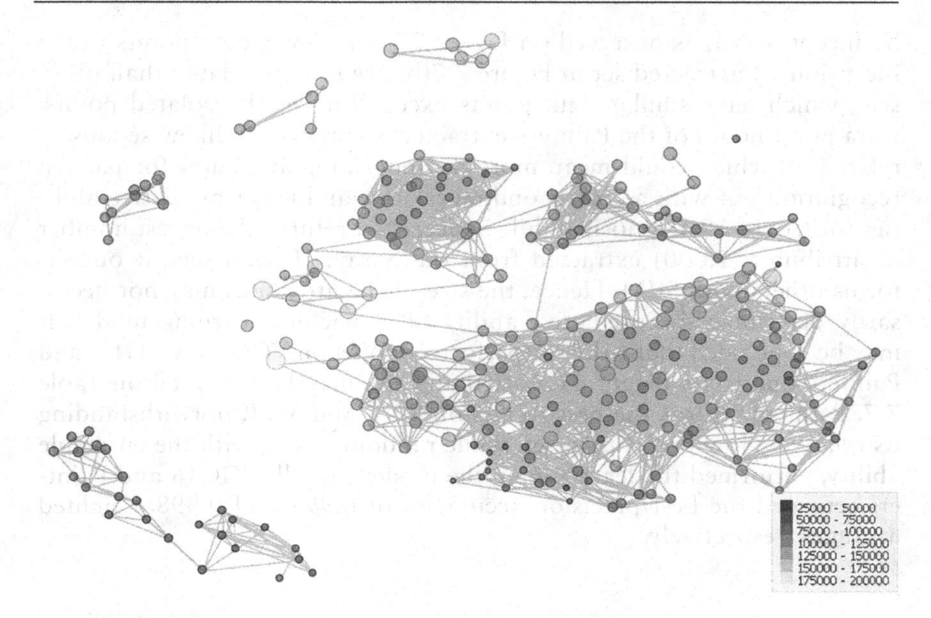

Figure 7.7(e) MDS plot for Painters.

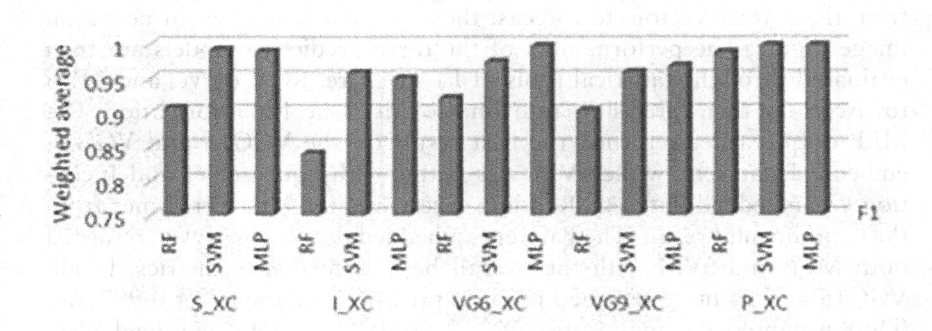

Figure 7.8 Plot of the F1 measure of the models on the five datasets.

v3 and Painters. It may be safe to mention SVM as the most suited for SqueezeNet in medical imaging, just as MLP deep learner has the overall best diagnosis ability across the five Covid-19 medical imaging sets. The level of similarities within attributes of the five datasets is plotted on the MDS. All the datasets have isolated data nodes, except for I_XC extracted by Inception v3, as observed on Figure 7.7. However, data points in the Inception v3 extracted set in Figure 7.7(b) are loosely related than other sets, which have similar data points except for clearly isolated points. Data point nodes of the Painters-extracted attributes are likewise loosely related (e) which could mean more distinguishing attributes for pattern recognition but with a corresponding relative and lesser predictive abilities for Covid-19 diagnosis. While SqueezeNet returns the fewest number of attributes (1,000) extracted from the X-ray:CT scan sets, it outperforms others with SVM. Hence, the size of the attributes may not necessarily determine the predictive ability of a machine learning model. It may be concluded that MLP outperformed SVM on VGG16, VGG19, and Painters due to its bigger data attributes. As may be observed on Table 7.7, Painters returns the best results on SVM and MLP, notwithstanding its relatively low pretraining set, while random forest, with the ensemble ability, performed the least among the models. In all, VGG16 and Painters returned the best precision accuracies of 0.997 and 0.998 weighted averages, respectively.

7.5 CONCLUSION AND RECOMMENDATION

The study deploys five deep image embedders to extract feature vectors from X-ray and CT scan image signals of both positive and negative cases of Covid-19, thereby employing three prominent machine learning algorithms to build an intelligent diagnostic model on patterns recognized from the feature vectors to forecast the coronavirus statuses of new scan image inputs. The performances of the three predictive models are then evaluated through statistical tools of F1 measure, ROC curve, and MDS to ascertain their precision rate on Covid-19 health informatics. The MLP deep learner returned the best result on the VGG16 and VGG19 embedded datasets, while SVM was better with SqueezeNet and Inception v3 embedded datasets. Random forest was the least performer from the experimental result. The Painters embedded dataset, however, returned both MLP and SVM with the overall best performance metrics. In all, VGG16 and Painters returned the best precision accuracies of 0.997 and 0.998 weighted averages, respectively. The MDS plot returned the dataset extracted by Painters as loosely networked data points, followed by Inception v3. Other embedders returned data point attributes with compact proximity, except for a few isolated cases. Future studies could further analyse the data points inherent in the embedded medical imaging sets towards implementing feature selection or principal component analysis. Information gain or ReliefF evaluators could calculate the weights of the data attributes to identify the most significant for the predictive analytics. Furthermore, while random forest is the least performer in this study, further study could be implemented with vote, bagging, and other ensembles to determine the predictive ability of ensembles on the predictive analytics of Covid-19 medical imaging.

ACKNOWLEDGEMENT

The authors appreciate the painstaking efforts of reviewers towards turning out a good study.

REFERENCES

[1] I. Toprak and O. Toygar, "Detection of spoofing attacks for ear biometrics through image quality assessment and deep learning," *Expert Systems with Applications*, 172, 2021.
[2] A. Abbas, M. Abdelsamea and M. Gaber, "Classification of covid-19 in chest x-ray images using detrac deep convolutional neural network," *arXiv:2003.13815*, 2020.
[3] T. Olaleye, O. Arogundade, C. Adenusi, S. Misra and A. Bello, "Evaluation of image filtering parameters for plant biometrics improvement using machine learning," in *icSoftComp*, Singapore, 2020.
[4] M. Tripathi, "Analysis of convolutional neural network based image classification techniques," *Journal of Innovative Image Processing (JIIP)*, pp. 100–117, 2021.
[5] T. Olaleye, O. Arogundade, A. Abayomi-Alli and K. Adesemowo, "An ensemble predictive analytics of COVID-19 infodemic tweets using bag of words," *Data Science for COVID-19*, pp. 365–380, 2021.
[6] WHO, "WHO coronavirus (COVID-19) dashboard," 4 February 2022. [Online]. Available: https://www.covid19.who.int.
[7] S. Sen, S. Saha, S. Chatterjee, S. Mirjalili and R. Sarkar, "A bi-stage feature selection approach for COVID-19 prediction using chest CT images," *Applied Intelligence*, 51(12), pp. 8985–9000, 2021.
[8] Shad, R., Cunningham, J. P., Ashley, E. A., Langlotz, C. P., & Hiesinger, W. (2021). Designing clinically translatable artificial intelligence systems for high-dimensional medical imaging. *Nature Machine Intelligence*, 3(11), 929–935.
[9] F. N. Iandola, S. Han, M. W. Moskewicz, K. Ashraf, W. J. Dally and K. Keutzer, "SqueezeNet: AlexNet-level accuracy with 50x fewer parameters and< 0.5 MB model size," *arXiv:1602.07360*, 2016.
[10] N. Dong and L. Z. H. W. F. Chang, "Inception v3 based cervical cell classification combined with artificially extracted features," *Applied Soft Computing*, 93(106311), 2020.
[11] N. C. Mithun, R. Panda, E. E. Papalexakis and A. K. Roy-Chowdhury, "Webly supervised joint embedding for cross-modal image-text retrieval," in *Proceedings of the 26th ACM International Conference on Multimedia*, 2018.
[12] S. Devadharshini, R. Kalaipriya, R. Rajmohan, M. Pavithra and T. Ananthkumar, "Performance Investigation of Hybrid YOLO-VGG16 Based Ship

Detection Framework Using SAR Images," in *In 2020 International Conference on System, Computation, Automa*, 2020.
[13] J. Zhang, Y. Xie, Y. Li, C. Shen and Y. Xia, "Covid-19 screening on chest x-ray images using deep learning based anomaly detection," *arXiv:2003.12338*, 2020.
[14] L. Wang, Z. Lin and A. Wong, "COVID-net: a tailored deep convolutional neural network design for detection of COVID19 cases from chest radiography images," *arXiv:2003.09871*, 2020.
[15] A. Narin, C. Kaya and Z. Pamuk, "Automatic detection of coronavirus disease (COVID-19) using x-ray images and deep convolutional neural networks," *arXiv:2003.10849*, 2020.
[16] P. Sethy and S. Behera, "Detection of coronavirus disease (COVID-19) based on deep features," *Preprints*, 2020.
[17] S. Niu, M. Liu, Y. Liu, J. Wang and H. Song, "Distant Domain Transfer Learning for Medical Imaging," *IEEE Journal of Biomedical Health Informatics*, 25(10), 2020.
[18] L. Li, L. Qin, Z. Xu, Y. Yin, X. Wang, B. Kong, J. Bai, Y. Lu, Z. Fang, Q. Song, K. Cao, D. Liu, G. Wang, Q. Xu, X. Fang, S. Zhang, J. Xia and J. Xia, "Using artificial intelligence to detect COVID-19 and community-acquired pneumonia based on pulmonary CT: Evaluation of the diagnostic accuracy," *Radiology*, pp. 1–8, 2020.
[19] R. Rehouma, M. Buchert and Y. P. Chen, "Machine learning for medical imaging-based COVID-19 detection and diagnosis," *International Journal of Intelligent Systems*, 36(9), pp. 5085–5115, 2021.
[20] S. R. Nayak, D. R. Nayak, U. Sinha, V. Arora and R. BilasPachori, "Application of deep learning techniques for detection of COVID-19 cases using chest X-ray images: A comprehensive study," *Biomedical Signal Processing and Control*, 64, 2021.
[21] F. Ucar and D. Korkmaz, "COVIDiagnosis-Net: Deep Bayes-SqueezeNet based diagnosis of the coronavirus disease 2019 (COVID-19) from X-ray images," *Medical Hypotheses*, 140(109761), 2020.
[22] A. Sedghi, H. Soheil, M. Alireza, Z. Erik and T. Clare, "Tesseract-medical imaging: open-source browser-based platform for artificial intelligence deployment in medical imaging," in *Medical Imaging 2019: Image-Guided Procedures, Robotic Interventions, and Modeling*, California, 10951, 2020.
[23] G.S. Fu, Y. Levin-Schwartz, Q.-H. Lin and D. Zhang, "Machine Learning for Medical Imaging," *Journal of Healthcare Engineering*, pp. 1–3, 2019.
[24] A. Ziller, D. Usynin, R. Braren, M. Makowsk, D. Rueckert and G. Kaissis, "Medical imaging deep learning with differential privacy," *Scientific Reports*, 11(1), pp. 1–8, 2021.
[25] S. Tabik, A. Gomez-Rios, J. Luis, Martin-Rodriquez and I. Sevillano-Garcia, "COVIDGR dataset and COVID-SDNet methodology for predicting COVID-19 based on chest X-ray images," *IEEE Journal of Biomedical and Health Informatics*, 24912, pp. 3595–3605, 2020.
[26] S. Minaee, R. Kafieh, M. Sonka, S. Yazdani and G. J. Soufi, "Deep-COVID: Predicting COVID-19 from chest X-ray images using deep transfer learning," *Medical Image Analysis*, 65, 2020.

[27] J. P. Cohen, L. Dao, K. Roth, P. Morrison, Y. Bengio and A. F, "Predicitng covid-19 pneumonia severity on chest x-ray with deep learning," *Cureus*, 12(7), 2020.
[28] M. Umer, I. Ashraf, S. Ullah, A. Mehmood and G. S. Choi, "COVINet: A convolutional neural network approach for predicting COVID-19 from chest X-ray images," *Journal of Ambient Intelligence and Humanized Computing*, 13(1), pp. 1–13, 2021.
[29] T. Rahman, "COVID-19 radiography database," 18 January 2022. [Online]. Available: https://www.kaggle.com/tawsifurrahman/covid19-radiography-database.
[30] P. Eduardo, "SARS-COV-2 Ct-scan dataset," 18 January 2022. [Online]. Available: https://www.kaggle.com/plameneduardo/sarscov2-ctscan-dataset.

Chapter 8

A comparative analysis for analysing the performance of convolutional neural network versus other machine learning techniques to assess cardiovascular disease

Harshavardhan Tiwari, Aishwarya M, Harshitha M V, Nida Shafin, Srushti J, and Tushita S

CONTENTS

8.1 Introduction 138
8.2 Related works 139
8.3 Methodology 142
8.3.1 Data collection 142
8.3.2 Data pre-processing 142
8.3.3 Training data 144
8.3.3.1 Convolutional neural network (CNN) algorithm 145
8.3.3.2 Decision tree 145
8.3.3.3 Random forest 145
8.3.3.4 Logistic regression 145
8.3.4 Model 145
8.3.5 Model classification 146
8.3.5.1 Coronary artery disease 146
8.3.5.2 Heart arrhythmias 146
8.3.5.3 Heart valve disease 146
8.3.5.4 Pericardial disease 146
8.4 Result 147
8.5 Conclusion 151
References 151

DOI: 10.1201/9781003305347-8

8.1 INTRODUCTION

The heart is the most important part of the human body [1]. It is the main organ that keeps all living beings or species alive, such as humans or animals. It pumps blood to every part of the human body [1]. The circulatory system or the cardiovascular system helps in supplying the body with blood. It consists of the heart, arteries, veins, and capillaries. Various other organs of the body are mainly dependent on the heart for their functioning. Suppose the heart fails to function properly, all other organs would also stop working, leading to cardiovascular diseases and danger. CVD is a very serious type of disease that is related to the anomalous condition of the heart, such as abnormal functioning of the right or left pulmonary arteries, pulmonary veins, valves, atriums, or ventricles which are part of the heart. Many concluded that cardiovascular diseases occur when the heart fails to pump blood as it should. Many precarious conditions of CVD encompass coronary heart disease (CHD), congenital cardiac disease, stroke, and many more. Heart diseases occur due to plenty of factors, such as aging, smoking, high consumption of alcohol, and also unhealthy lifestyle [2]. Other risk factors include high blood pressure, cholesterol, obesity, use of tobacco, and bad nutrition [9]. One of the preventive and easiest ways to control heart disease is to balance or entirely modify all the risk factors that lead to the disease, which will decrease the chances of getting cardiac disease. It has many symptoms, such as high heart rate, difficulty in breathing, and chest pain. At present, 85% of deaths worldwide occur due to cardiac arrest.

Per the research and update from WHO, cardiovascular disease is the primary cause of death all over the world, with an estimate of around 17.9 million lives each year [2,4]. Therefore, it has become indispensable to lead a healthy lifestyle and to detect heart-related diseases immediately to avoid death. The medical industries or the healthcare organizations are facing barriers in the determination of cardiovascular disease at the early stage. It cannot be predicted so easily because it is a crucial task that will demand an enormous amount of knowledge and skilled expertise. A good data-driven system for the prediction of cardiovascular diseases can entirely improve the prevention process. Diagnosing the disease at the earliest stage can reduce the risk of death in many cases. Machine learning (ML) and artificial intelligence (AI) hold a major part in many fields, especially in the medical industry [2]. AI, ML, and DL are currently becoming a reason for development in the medical industry by supporting various technologies which rely on data mining [4,5]. Mainly, ML algorithms are known for playing a very important role in the prediction of heart diseases. ML algorithms help solve real-life problems automatically. This work deals with the prediction and classification of cardiovascular diseases based on ML and DL algorithms, which slightly reduce the risk of death due to cardiac disease [3]. This work can be utilized to help specialists in the medical field by procuring opinions in many critical situations that can save the lives of patients.

Various ML algorithms, such as random forest, decision tree, and logistic regression, are taken into consideration to predict whether CVD is present or not in a particular person [6]. The CVD prediction and classification system consists of training and test datasets. The fields considered in the dataset are gender, age, smoking habits, BP medication, prevalent stroke, hypertension, cholesterol, BMI, heart rate, glucose, and so on. Supervised learning algorithms analyse the training data and deduce a function from the labelled training dataset. The output will be seen as binary values that will predict whether the patient has heart disease. Later, the model will be compared with the other conventional machine learning algorithms. A deep learning algorithm, such as the CNN algorithm, is used to predict and classify diseases [5]. The different types of cardiovascular diseases include coronary artery disease, heart arrhythmia, heart valve disease, and pericardial disease. To classify different types of diseases, we make use of the CNN algorithm [7]. It also helps in enhancing and improving the diagnosis of heart disease which helps in efficient and quick decision-making to save the lives of many [10].

8.2 RELATED WORKS

V. V. Ramalingham et al. (2018) [1]: The study here introduces a detailed analysis of numerous features based on algorithms and strategies to research their performance. The models are totally based on supervised studying algorithms and their rules; the algorithms are k-nearest neighbors, decision trees, random forests, support vector machines, naïve Bayes, and ensemble models, which are very famous to various researchers. Alternating the decision trees whilst used together with PCA has accomplished extraordinary property; however, decision trees have failed to perform in some different instances which might be used for overfitting. The random forest and ensemble models have been done properly due to the fact that they remedy the inconvenience of overfitting by utilizing multiple algorithms. Models absolutely established on the naïve Bayes classifier were swift and have additionally executed properly. SVM executed exceptionally nice for the maximum of given cases. Structures which thoroughly relied on machine learning algorithms and their strategies have been very accurate in forecasting heart-associated illness.

Devansh Shah et al. (2020) [2]: This paper gives diverse traits related to cardiac disorder and the model in support of supervised learning algorithms together with decision trees, naïve Bayes, k-nearest neighbors, and random forest algorithms. It employs a dataset from the database of the UCI repository. The dataset accommodates 303 instances and 76 attributes; of these 76 attributes, the most effective 14 attributes are taken into consideration for checking out. This paper targets to ascertain the chance of developing coronary heart ailment in patients. The result portrays that the inflated accuracy score is done with the K-NN set of rules.

M Swamy et al. (2020) [3]: The paper is prepared with a threefold class and data mining strategies for CVD, machine learning, and deep learning models for CVD prediction. The execution metrics used for finding the accuracy, the tools used for each category of these strategies, and the dataset used for prediction and classification are also stated and compiled within this particular paper survey.

Rohit Bharti et al. (2021) [4]: This paper includes distinct machine learning and deep learning algorithms which were carried out by evaluating the consequences and evaluation of the UCI machine learning to know coronary heart disorder records. These records comprise 14 attributes used to represent the analysis. Varieties of encouraging results are completed and then they had been tested employing the accuracy and the confusion matrix. Then the dataset includes a few immaterial capabilities that are grasped using isolation forest, and statistics are also modified to get higher outcomes. And the way this takes a look may be mixed with a few multimedia technology; for example, mobile gadgets are also mentioned. With the usage of the deep learning technique, 94.2% accuracy was acquired.

Kishore G R et al. (2021) [5]: Cardiac arrhythmia is a circumstance wherein the electric rhythm of the coronary heart is peculiar. The electric rhythm could be very irregular leading to the disruption of the cardiac rhythm. In this work, an efficient arrhythmia-type version was used by leveraging a 2D convolutional neural network. This work is completely based on the unsupervised learning-based technique, that is, we've bypassed the traditional data pre-processing, characteristic extraction, and so on, but the completed accuracy is in the direction of 90% and more or much less the same degree of sensitivity. Additionally, they endorsed a technique to classify heartbeats which helps in the classification of five special classes of arrhythmia that are compliant with the AAMI EC57 general. This approach is primarily based on the t-SNE model and has been capable of obtaining an average accuracy of 93%.

Youness Khourdifi et al. (2018) [6]: It employed a dataset from the UCI Machine Learning Repository and became skilled in five exceptional machine learning algorithms; they are the support vector machine, naïve Bayes, random forest, k-nearest neighbor, and artificial neural network algorithms. Then it was examined using the test information and the confusion matrix wherein the output of the KNN and the RF algorithm was the most accurate, and the evaluation section revealed the success of hybrid PSO and ACO strategies in disorder prognosis compared with other present processes.

Christian Castillo et al. (2020) [7]: They accumulated records through Media Norte Hospital in Mexico which consists of 800 data and 141 signs, together with age, blood pressure rate, weight, glucose, and clinical symptoms. The scattering of the accumulated data was not balanced, wherein 17% of its data had hypertension, 16% consisted of an ischemic coronary heart disorder, around 7% of the information had blended heart disorder, and the rest of the 8% information consisted of a valvular coronary heart ailment. Experiments had been carried out with bidirectional and unidirectional neural network communities. Results and model indicated that an ensemble classifier with BIGRU or BILSTM model, together with a CNN model, had first-class category overall outcome of accuracy and F1 score between 91% and 96% for the specific varieties of heart disorder. Those outcomes are aggressive and assuring for coronary heart disorder records. They confirmed that the ensembles learning framework primarily gleaned from deep learning models. The intended structure can have the most accurate models which can be modified for scientific information and analysis.

Abdulziz Albahr et al. (2021) [8]: In this research, the prediction model is lodged in a new regulation centred totally on deteriorating the weight in accordance with load matrices predictable error and evaluating the outcome in opposition to its parent (RSD-ANN). The overall outcome of RSD-ANN is a ways ahead of the existing methods. The overall accuracy obtained is 96.30%, based totally on the experiments with the use of a 10-fold cross-validation or holdout method.

Keshav Srivastava et al. (2020) [9]: This paper proposed a unique approach to determining the disorder by the usage of the coronary heart disease dataset by integrating the computational power of various machine learning and deep learning algorithms; and the conclusion is that among all the algorithms, k-nearest neighbors offer the most accurate at 87%. In the company, the latest is a web app using Flask in Python, wherein the person (user) can input the attributes and predict the coronary heart disorder, whether the person has coronary heart ailment or not, with the model that is skilled in the usage of the dataset.

D. Komalavalli et al. (2021) [10]: In this paper, there are many techniques, like clustering, type, genetic algorithm (GA), neural network (NN), decision support systems (DSS), and many more, to find out the severity of the affected person's medical records. Primarily based on the fuzzy rule, the machine can be used for predicting coronary heart disease. A weighted fuzzy rule–based clinical decision support system (CDSS) is presented to analyse coronary artery disease. The UCI repository routinely accumulates the statistics from the person's clinical dataset and uses fuzzy guidelines. As a primary stage, the data pre-processing is implemented to throw away the missing data and the unwanted data from the clinical dataset. Then the machine is used to define the weighted fuzzy policies and function extraction. A choice of regulations is built and is put into the database to identify the prevalence model, which is based on the frequency stage; the burden is premeditated for every decision rule received from the weighted fuzzy policies, and training datasets are received with the support of the fuzzy integration function. Then the inputs are given to the developed system to categorize and expect coronary heart diseases.

8.3 METHODOLOGY

The workflow methodology (Figure 8.1) gives the overview of the development of the model which is used to predict cardiovascular disease using the various datasets which was collected by the UCI Machine Learning Repository and the detailed and step-by-step process for the development of the model.

8.3.1 Data collection

The information has been accumulated from the UCI Machine Learning Repository [9,10]. This record is mentioned as the heart disease dataset that's raw information inclusive of 14 attributes (Figure 8.2). The UCI Machine Learning Repository includes N wide variety of datasets from nearly many domains where these statistics are particularly utilized in ML [10]. numerous research had been conducted by the usage of the datasets from the repository. The repository was created by David Aha and fellow students of UCI Irvine in 1987.

The coronary heart disease dataset used here consists of records from four institutes:

- Hungarian Institute of Cardiology, Budapest
- University Hospital, Zurich, Switzerland
- University Hospital, Basel, Switzerland
- VA Medical Center, Long Beach, and Cleveland Clinic Foundation

8.3.2 Data pre-processing

Cardiovascular disorder data is pre-processed after gathering the information (Figure 8.3). The dataset consists of 4,241 records of information; there are some lacking values. Data pre-processing also entails doing away with that dataset, and the ultimate 3,752 datasets go through additional processes. The binary classifications and multiclass variables are added for the accrediting of the dataset [10]. The multiclass variable is used to test whether the patient has been affected by cardiovascular disease or not. In the dataset of the patient having cardiovascular ailment, the value is 1, else the value is 0, which shows the presence and absence of cardiovascular disorder in the patient, and the information is classified as 1, 2, 3, and 4 based on the dataset: they are 1, coronary artery disease; 2, heart arrhythmia; 3, heart valve disease; and 4, pericardial disease. The pre-processing of data is performed by converting medical records into diagnosis values. The data after the pre-processing for the 4,241 patient dataset indicates that 644 records have the value of 1, 2, 3, and 4, showing the presence of cardiovascular disease; while the remaining 3,596 have the value of 0, indicating the absence of cardiovascular disease.

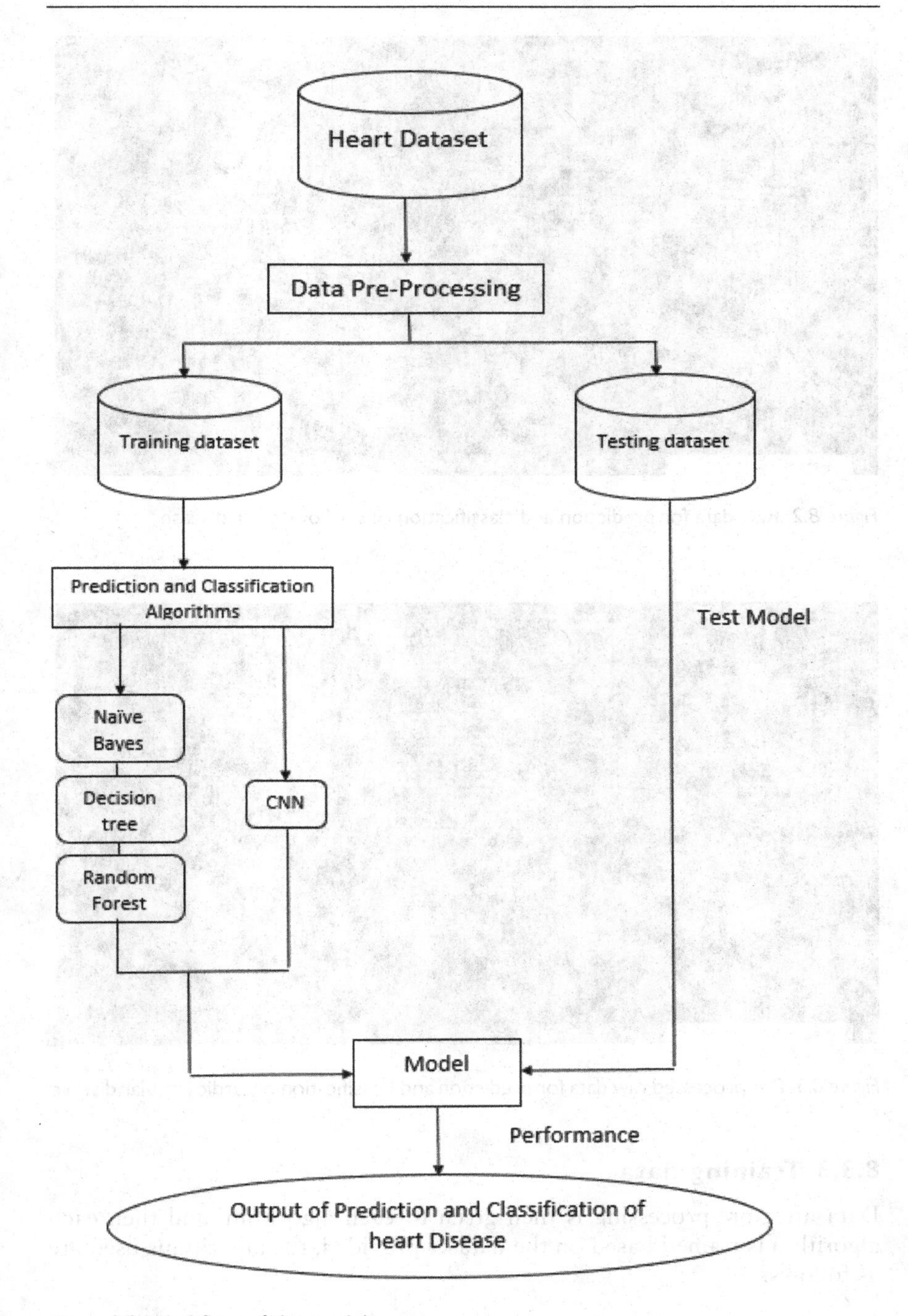

Figure 8.1 Workflow of the model.

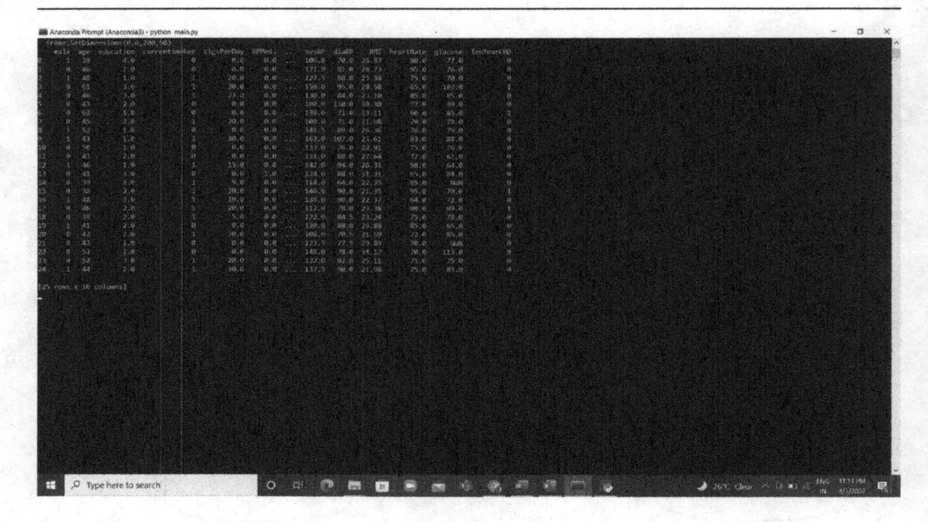

Figure 8.2 Raw data for prediction and classification of cardiovascular disease.

Figure 8.3 Pre-processed raw data for prediction and classification of cardiovascular disease.

8.3.3 Training data

Data after pre-processing is then given to each algorithm, and then each algorithm is trained based on the dataset provided; the algorithms used are as follows:

8.3.3.1 Convolutional neural network (CNN) algorithm

The task of identifying the presence of heart disease in patients refers back to the assignment of binary classification. Underneath some repression, the convolutional neural network has been verified to be an effective and constant classifier under operated settings.

8.3.3.2 Decision tree

The decision tree is a class that works on specific as well as numerical statistics. A decision tree is used to create structures similar to a tree [2]. It is straightforward and extensively used to address clinical datasets. It is simple to implement and examine the statistics inside the tree-formed graph. The decision tree model makes analysis primarily based on the tree nodes [2]. The set of rules divides the statistics into two and more parallel and related sets based on the maximum dominant signs.

8.3.3.3 Random forest

Random forest is a supervised form of an algorithmic approach [1]. In this algorithm, numerous trees create a forest. Where every tree in a random forest conceals a category presumption, the magnificence with the maximum referendum will become a model's prognosticate. It is used for the category in addition to regression responsibilities; however, it can do properly with classification responsibilities and can output final results lacking values [6].

8.3.3.4 Logistic regression

The logistic regression is used as a classification set rule often for binary category problems. As opposed to the fitting of a hyperplane or straight line, the logistic regression set of rules uses the logistic feature to provide the result of a linear equation between 0 and 1 [9].

8.3.4 Model

Our model is trained with 4,241 datasets; then the model is tested with the test data to find out whether the patient has cardiovascular disease or not; then the accuracy of each algorithm is calculated by entering the test data to predict and classify the cardiovascular disease separately, and the graph is plotted to check the accuracy of each algorithm which is the final output of the model.

8.3.5 Model classification

The assembly of datasets is accomplished based on the variables and basis of the decision tree (DT) features. After that, the classifiers are implemented to every assembled dataset to estimate its overall performance. The excellent results of models are detected from the previous results based totally on their charge of mistakes. The result is further optimized by choosing the decision tree to assemble with a high charge of mistakes and extraction of its classifier capabilities. The overall production of the classifier is evaluated for mistakes escalation on the dataset.

The diseases are classified into four types. A brief description of each is given as follows:

8.3.5.1 Coronary artery disease

The accumulation of a fat layer called plaque in the coronary arteries is responsible for diseases such as coronary artery disease, also known as ischemic heart disease or coronary heart disease. The difficulty in delivering oxygen-rich blood to the body, chest pain, shortness of breath, fatigue, dizziness, or discomfort in the shoulders or arms will let us know the symptoms of coronary artery disease. Different tests, such as echocardiogram, cardiac catheterization, and EKG, may be included for detecting the disease [7,10].

8.3.5.2 Heart arrhythmias

Arrhythmia can be understood as an abnormal heartbeat in the human body. As the disease is concerned with the heartbeat, one of the important symptoms is a palpitations and skipped heartbeat. EKG, EPS, tilt table test, stress test, ambulatory monitors, echocardiogram, and cardiac catheterization are conducted to detect these diseases [5].

8.3.5.3 Heart valve disease

The improper working of one or more valves in the heart will give rise to heart valve disease. Not detecting this type of disease at an early stage might lead to complications, such as heart failure, blood clots, stroke, heart rhythm abnormalities, and death. The detection of such diseases also gives a chance for surgeries to rebuild the heart valve or to replace it.

8.3.5.4 Pericardial disease

The increase in the pericardial fluid is called pericardial disease. It leads to swelling of the pericardium (tissue surrounding the heart). This kind of disease is usually hard to determine because a perfect cause for the disease may not be found. Infections, such as COVID-19, also is a cause of pericardial disease. The risk of pericardial disease is reduced if proper treatment is given at a proper time and if the disease is detected at the earliest.

8.4 RESULT

The model we developed will produce the performance of the individual algorithm and produce the graph of each algorithm that we selected, as shown in Figure 8.4.

The figure shows the accuracy, mean absolute error (MAE), root mean squared error (RMSE), performance of the linear regression (R-squared), and mean squared error (MSE) values of all the algorithms.

After analysing the accuracy of each algorithm, we have used the convolutional neural network (CNN) algorithm to predict and classify the cardiovascular disease. If the disease is present in a particular patient (Figure 8.5), it is classified into four types: coronary artery disease, pericardial disease, heart arrhythmia, and heart valve disease. Then it displays the type of cardiovascular disease present in the person. If there is no disease, then it displays as no disease in a message box format.

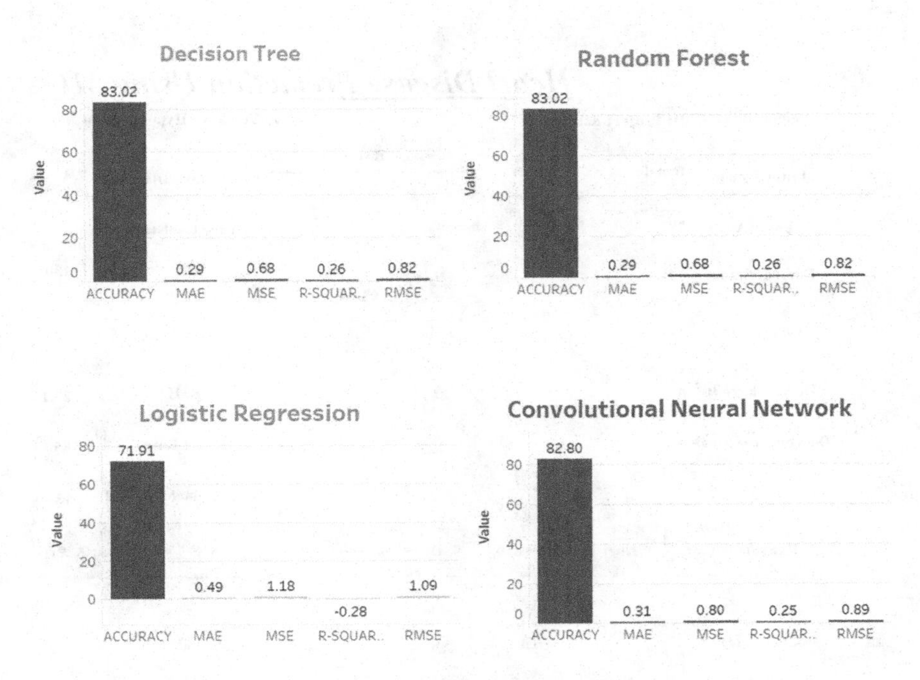

Figure 8.4 Individual performance of algorithms in predicting cardiovascular disease.

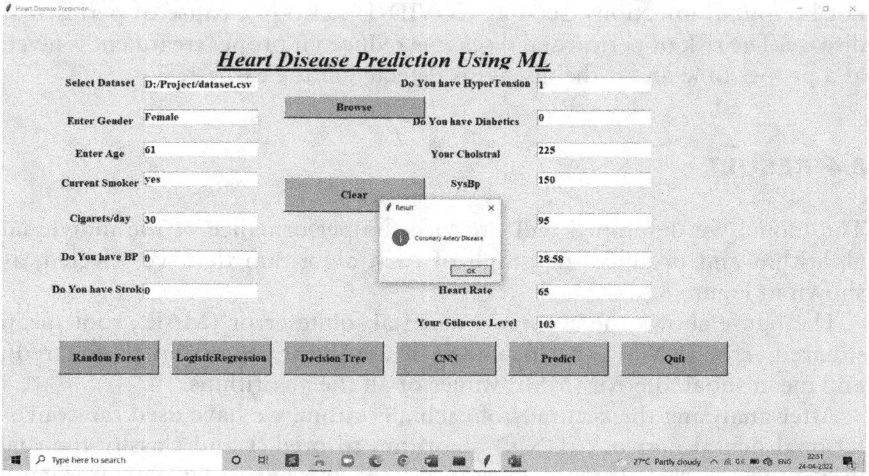

Heart Disease Prediction

Heart Disease Prediction Using ML

Select Dataset D:/Project/dataset.csv
Browse
Do You have HyperTension 1
Enter Gender Female
Do You have Diabetics 0
Enter Age 61
Your Cholstral 225
Current Smoker yes
SysBp 150
Clear
Cigarets/day 30
95
Do You have BP 0
28.58
Do You have Strok 0
Heart Rate 65
Your Gulucose Level 103

Result
Coronary Artery Disease
OK

Random Forest | LogisticRegression | Decision Tree | CNN | Predict | Quit

Type here to search
27°C Partly cloudy ENG 22:51 24-04-2022

Figure 8.5 Coronary artery disease present and predicted.

Heart Disease Prediction

Heart Disease Prediction Using ML

Select Dataset D:/Project/dataset.csv
Do You have HyperTension 0
Browse
Enter Gender female
Do You have Diabetics 1
Enter Age 46
Your Cholstral 285
Current Smoker yes
130
Clear
Cigarets/day 23
84
Do You have BP 0
BMI 23.1
Do You have Strok 0
Heart Rate 85
Your Gulucose Level 130

Result
Heart Valve Disease
OK

Random Forest | LogisticRegression | Decision Tree | CNN | Pre

Type here to search

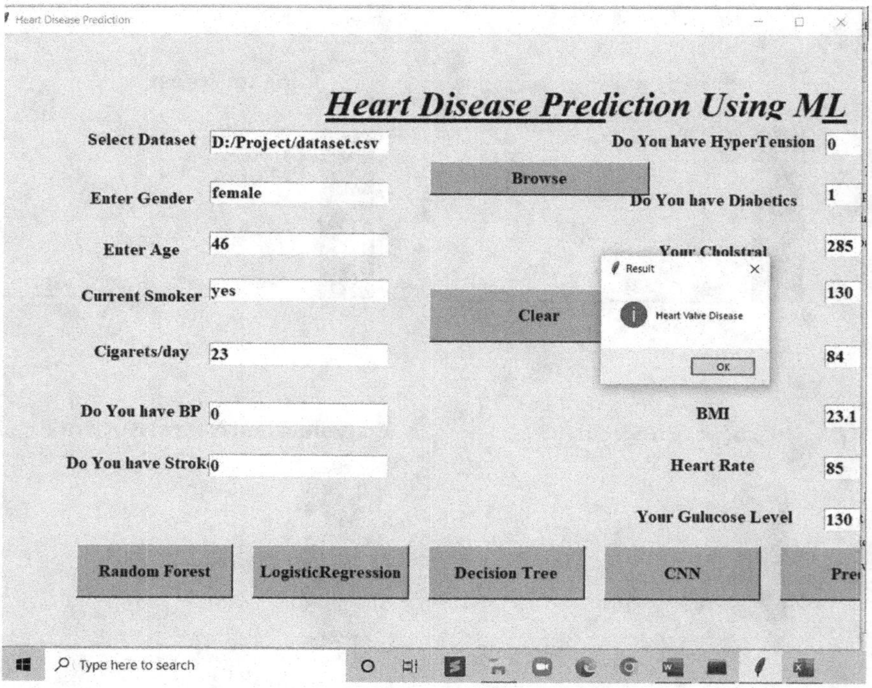

Figure 8.6 Heart valve disease present and predicted.

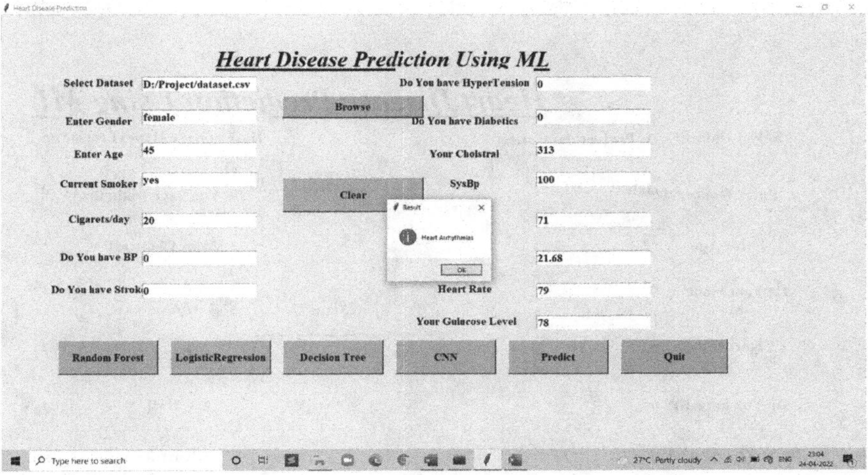

Figure 8.7 Heart arrhythmias present and predicted.

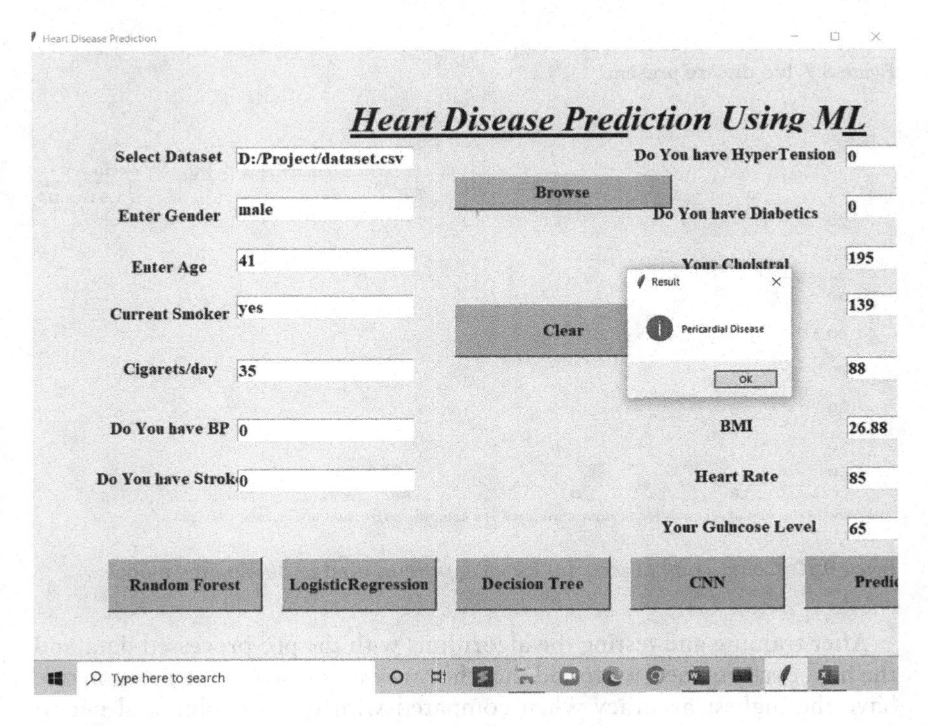

Figure 8.8 Pericardial disease present and predicted.

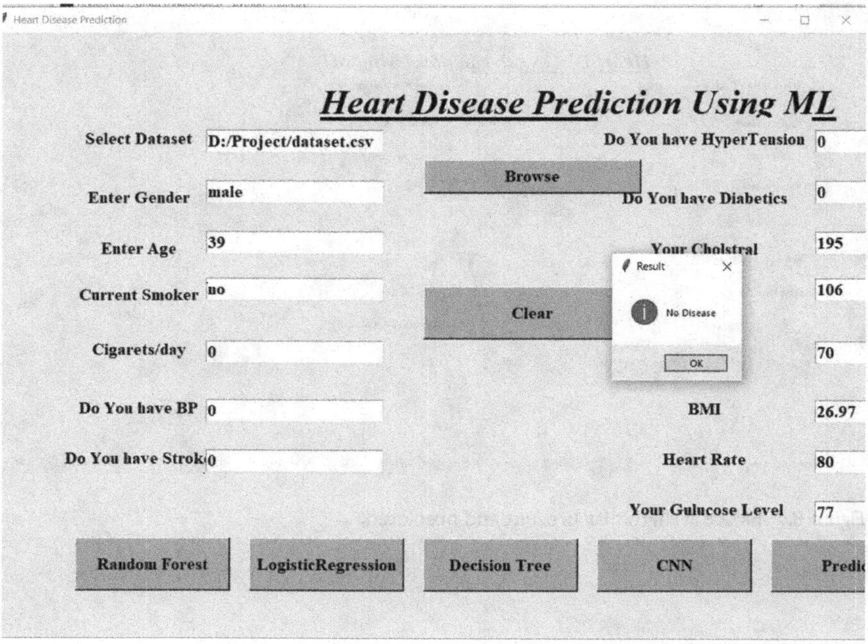

Figure 8.9 No disease present.

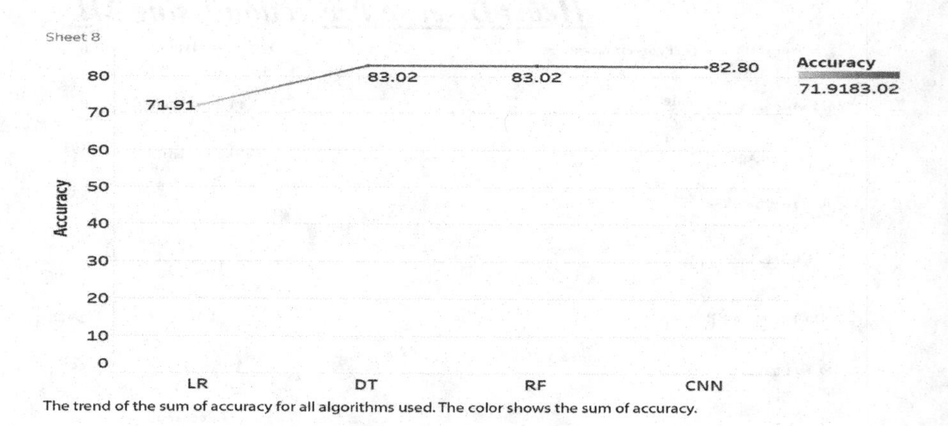

Figure 8.10 Comparison of accuracy for all algorithms used to develop the model.

After training and testing the algorithms with the pre-processed data and the final performance, we found that the random forest and the decision tree have the highest accuracy when compared with the convolutional neural network; and then the logistic regression is less accurate compared to the convolutional neural network. Figure 8.10 shows the graph which displays the accuracy of all the algorithms used in this work.

8.5 CONCLUSION

With the increasing number of deaths due to cardiac diseases, it is an obligation to develop a model to predict coronary heart diseases effectively and approximately. This particular work helped develop an automated system, which will facilitate the doctor to have patient health records and generate e-prescriptions using data mining techniques. The agenda of this project was to compare CNN algorithm with other ML algorithms for detecting heart diseases. It mainly compares various algorithms used for predicting heart diseases using the ML repository dataset. Various data is collected and processed, and the model is trained with convolutional neural network (CNN) algorithm to classify different cardiovascular diseases. The experimental results illustrate that the algorithms achieve the highest accuracy and, thus, successfully improve the prediction accuracy. Further, this work aims at reducing the risks of heart diseases by predicting them in an early stage. The final output is achieved through the 13 attributes in the dataset and 1 labelled class which will help the doctors to predict the cardiovascular disease in the earlier stage; then the patients can be treated based on the output produced which helps to reduce the death rate due to cardiovascular disease. We have not yet achieved the best accuracy in this work but can get the highest accuracy by increasing the number of datasets. As we train the models with many datasets, the output produced would be more accurate in the exact results. In our work, we have proved that the ML algorithm works better when compared with the DL algorithm, but if we train 100% of the data that is required for the model, we'll get the best results.

REFERENCES

[1] V.V. Ramalingam, Ayantan Dandapath, and M. Karthik Raja, (2018), "Heart disease prediction using machine learning techniques – A Survey", *International Journal of Engineering & Technology*, 684–687, 2019 IEEE Paper.

[2] Devansh Shah, Santosh Kumar Bharti, and Samir Patel, (2020), *Heart disease prediction using machine learning techniques*. Springer Nature Singapore Pvt Ltd.

[3] Swathy, M., & Saruladha, K. (2022). A comparative study of classification and prediction of Cardio-Vascular Diseases (CVD) using Machine Learning and Deep Learning techniques. *ICT Express*, 8(1), 109–116.

[4] Bharti, R., Khamparia, A., Shabaz, M., Dhiman, G., Pande, S., & Singh, P. (2021). Prediction of heart disease using a combination of machine learning

and deep learning. *Computational intelligence and neuroscience*, vol. 2021, Article ID 8387680, 11 pages. https://doi.org/10.1155/2021/8387680

[5] G.R. Kishore, and B.R. Shubhamangala, (2021), "Heartbeat classification and arrhythmia detection using deep learning," *Turkish Journal of Computer and Mathematics Education*, 12(7), 1457–1464, Research Article.

[6] Youness Khourdifi, and Mohamed Bahaj, (2019), "Heart disease prediction and classification using machine learning algorithms optimized by particle swarm optimization and ant colony optimization," *International Journal of Intelligent Engineering and Systems*, 12(1). DOI: 10.22266/ijies2019.0228.24

[7] Asma Baccouche, Begonya Garcia-Zapirain, Cristian Castillo Olea, and Adel Elmaghraby, (2020), "Ensemble deep learning models for heart disease classification: A case study from Mexico," *Information*, 11, 207.

[8] Albahr, A., Albahar, M., Thanoon, M., & Binsawad, M. (2021). Computational learning model for prediction of heart disease using machine learning based on a new regularizer. *Computational Intelligence and Neuroscience*, vol. 2021, Article ID 8628335, 10 pages. https://doi.org/10.1155/2021/8628335

[9] Dilip Kumar Choubey, and Keshav Srivastava, (2020, May), "Heart disease prediction using machine learning, and data mining," *International Journal of Recent Technology and Engineering (IJRTE)*, 9(1), ISSN: 2277-3878.

[10] D. Komalavalli, R. Sangeethapriya, R. Indhu, N. Kanimozhi, and G. Kasthuri, "An effective heart disease prediction using machine learning technique," *ICTACT Journal on Soft Computing*, 11(03), April 2021, pp. 2323–2327. DOI: 10.21917/ijsc.2021.0332.

Chapter 9

A study of machine learning initiatives in the global healthcare sector using different case studies

Swati Singh

CONTENTS

9.1 Introduction 153
9.2 Global healthcare sector and machine learning 156
9.3 Decision support system (DSS): A need for the healthcare sector 157
9.4 Decision support based on technology 158
9.5 Decision support system (DSS) in healthcare 158
9.6 Various machine learning methods used for disease prediction 158
9.7 Conclusion and future scope 160
References 161

9.1 INTRODUCTION

Healthcare is currently one of the fastest growing sectors and is now on the cusp of a complete worldwide transformation [1], [2]. Many groups are dealing with a huge volume of datasets which have to be identified with machine learning [3]. By collecting insights from these data, they are competent to work organized and more proficiently as well as gain a lead over their challengers. Advanced analytical models with machine learning algorithms have been successfully implemented in multiple domains [4], [5] (Figure 9.1).

Machine learning methods are applied in day-to-day activities, such as searching, advertising, YouTube or Google. Today, human healthcare information, an interdisciplinary field, has become identical with advanced expansion and data management competitions [5]. Medicine or health informatics is a systematic field that provides knowledge of medical information, data storage, retrieval, optimal use and decision-making for problem-solving [6], [7]. Advances in information gathering, treatment, communication and research have led to the enormous development of technology in health

DOI: 10.1201/9781003305347-9

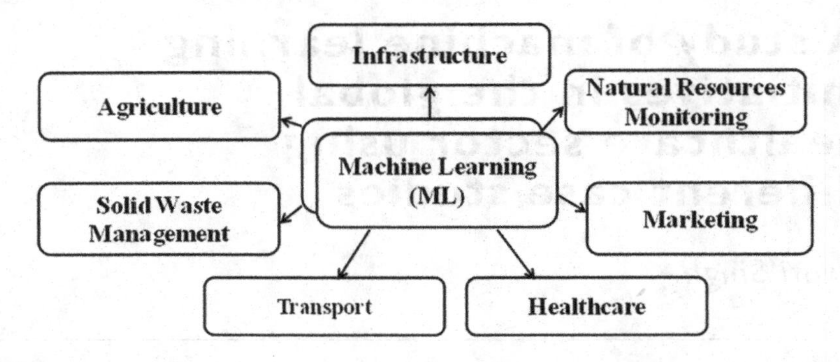

Figure 9.1 Use of machine learning in different fields.

over the years [8]–[10]. We can define 'machine learning' as a broad multidisciplinary field with backgrounds in algebra, data processing, statistics, information analysis and others, which makes it difficult to come back with an innovative description [11]. ML is a special technique of artificial intelligence (AI) that assembles important information through training datasets [12]. In the process, we're not giving orders to the machines where to look, and it's on the basis of a tree with lots of living or dead divisions. Machine learning is divided into the following four classes shown in Figure 9.2.

Supervised learning (SL) is machine learning in which the task of learning is based on training and algorithm development application that matches the inputs with the respective outputs [13] (Figure 9.3). In this learning process, a simple machine learning model is separated into two phases: (i) training and (ii) testing [14], [15]. In the training phase, samples are taken as input in the training data in which the learning algorithm or features are learned by the learner and the learning model is constructed [16]. And in the testing phase, the learning model exercises the execution engine to make likelihoods for the test or manufactured data. Tagged data is the major output of the learning model which provides the absolute anticipated or classified data [17].

A major difference between supervised ML and unsupervised ML is the problems the which the final models are used to solve [18]. Both models learn from training datasets, but the strengths of each method lie in dissimilar applications. Supervised learning will learn the association between input and output, concluded using labelled training datasets; therefore, it is used to categorize new data using these learned arrays or in forecasting outputs [19]. Unsupervised machine learning, however, is advantageous in ruling primary patterns and connections within unlabelled, raw datasets [20]. This makes it mainly suitable for tentative data analysis, clustering, or segmentation of datasets, or projects to understand how data types connect to other features for mechanized approval systems. Unsupervised machine

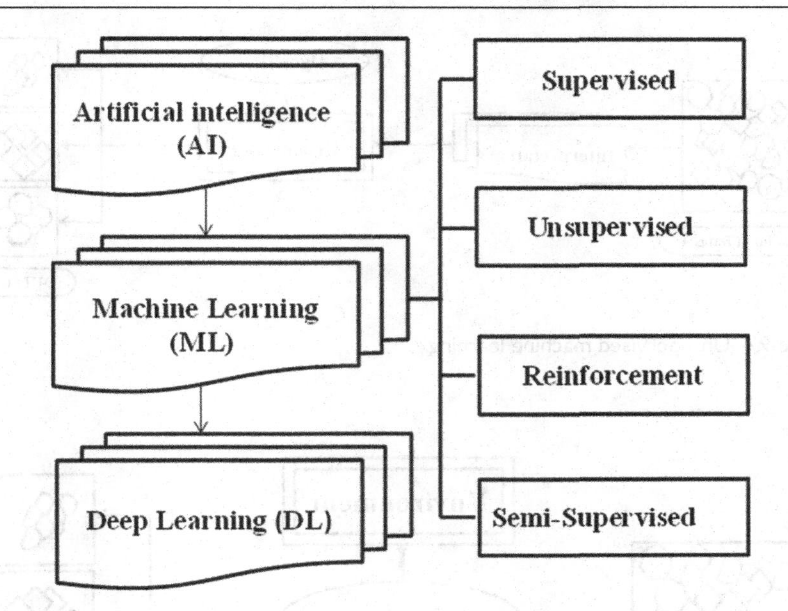

Figure 9.2 Machine learning classification.

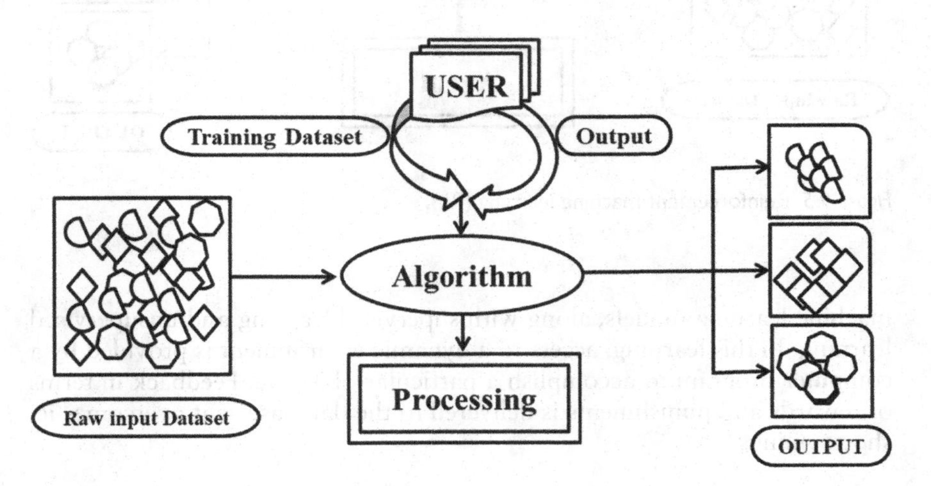

Figure 9.3 Mechanism of supervised learning (SL).

learning is a type of machine learning that acquires arrays from untagged datasets (Figure 9.4).

Reinforcement machine learning (RL) is a field of machine learning that compacts with how intelligent superiors must perform arrangements in an environment to make best use of the perception of a cumulative reward [21]–[23] (Figure 9.5). Reinforcement learning is one of the three basic

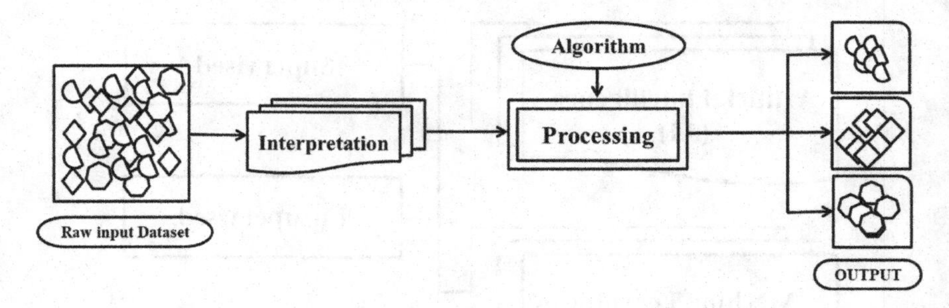

Figure 9.4 Unsupervised machine learning.

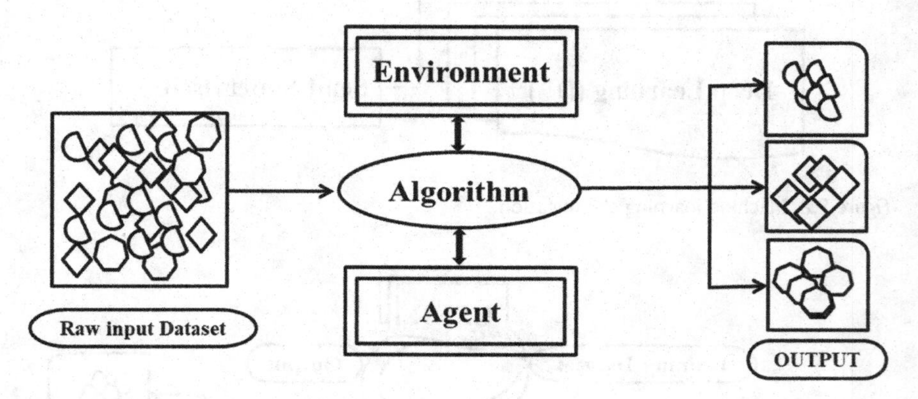

Figure 9.5 Reinforcement machine learning (RL).

machine learning models, along with supervised learning and unsupervised learning. In this learning, access to a dynamic environment is provided by a computer program to accomplish a particular objective. Feedback in terms of rewards and punishments is delivered to the database as it overcomes its shortcomings.

9.2 GLOBAL HEALTHCARE SECTOR AND MACHINE LEARNING

The alarming upgrading of health information provided new prospects for patient recovery; machine learning demonstrates an important performance in healthcare and is mostly applied to healthcare, including computer-aided diagnostics, image annotation, image registration and image database [24], [25].

Images directed medical support, image retrieval, multimodal image fusion and medical image segmentation, where the deficiency may be eternal [26]. Machine learning has possibly partial societal effects in the healthcare sector [27]. Machine learning offers solutions to reduce the growing cost of healthcare and build an improved communication between patient and doctor. ML solutions will be used for a heterogeneity of health-related practices; some include servicing physicians, identifying lots of personalized medicaments and therapies for patient, as well as serving patients once identified and if they must record follow-up schedules [28], [29].

Presently, a massive amount of information has become available in the healthcare sector. This includes EMR which includes information that may be either unstructured or structured [30]. Structured health information is dataset that is easy to examine in a record and is not limited to statistics and modules as well as patient weight, and will even carry common symptoms, such as headache and abdominal pain [31]. The majority of medical information is amorphous information contained within a number of numerous notes, images, audio and videos, reports, and release outlines [32]. It is very tedious to investigate and measure the interaction between provider and patient. The discussion is extremely personal and can take many alternate instructions [33].

Machine learning algorithms are useful in recognizing complex arrays within rich and massive data [34]. This feature is particularly suitable for clinical uses, particularly those that depend on advanced proteomics and genomics-based magnitudes [35]. It is often used in the diagnosis and detection of many diseases [36]. In medical uses, machine learning procedures will form advanced judgements concerning treatment procedures for patients by suggesting suitable healthcare systems to implement [37]. Healthcare management is using this system to predict waiting intervals for patients waiting for locations in the emergency division. These models use elements such as patient information, level of discomfort, urgent department charts and even hospital room layout to eliminate their waiting times [38], [39]. Using prehistoric models, clinics will think of hospital room charges as helping patients by reducing prices, increasing accuracy or channelling the involvement into a concise offer [40]–[42].

9.3 DECISION SUPPORT SYSTEM (DSS): A NEED FOR THE HEALTHCARE SECTOR

Errors in the healthcare system kill the majority of global residents each year, and thousands more suffer non-fatal burns from continuing causes. The Health Information Technology (IT) Framework suggested some approaches, such as associating, knowing the consumer choice of clinicians and organizations, and adopting IT [43].

9.4 DECISION SUPPORT BASED ON TECHNOLOGY

An effective machine learning–based healthcare system capitalizes on the huge computing proficiency of the computer and, similarly, the intellectual aptitude of the doctor [44]. Both the machine and the doctor are inspecting for outlines, but the doctor may or may not be aware of the specifics of each disease. The machine will perform all these responsibilities and then deliver its results to the doctor for validation [45].

9.5 DECISION SUPPORT SYSTEM (DSS) IN HEALTHCARE

The DSS benefits the confidence arm of the health centre, having tracking of charges, financial records, outlay and details of payable accounts. This process also supports the maintenance of the patient assurance policy and other repayment options [46]. It is one such organization which gives entirely changed modules for DSS in healthcare. Decision support systems support doctors in conducting disease screenings by arranging perceptions of health issues or by uncovering background knowledge about single patients. It also helps in recognizing the patient's condition and an attendant that suggests the patient to use the appropriate medication at a particular time through a web-based structure linked to computerized patient/medical records [47], [48].

9.6 VARIOUS MACHINE LEARNING METHODS USED FOR DISEASE PREDICTION

Nowadays, ML methods are frequently used to increase the accuracy of several disease diagnoses (Table 9.1). Machine learning systems are being used to forecast various diseases. Several kinds of classification algorithms, such as support vector machines (SVM) and decision trees, are used; and datasets are widely deliberated for prediction. The application of machine learning algorithm advancement in the healthcare sector is very helpful. Medical services need to be upgraded so that improved decisions can be made for disease analysis and treatment-related decisions. Machine learning in healthcare enables persons to practice vast and difficult medical datasets and then examine them with clinical assumptions. It can then be used by physicians in terms of medical care. Therefore, patient approval can increase when machine learning is implemented in healthcare.

ML methods are frequently used to increase the accuracy of diagnosis in heart disease. The dataset from the UCI Machine Learning Repository [64] projected an ML algorithm for heart disease detection and analysis using neural network analysis. The author stated that a parallel method is also accepted in the training segment to increase the efficiency of the classification

Table 9.1 Machine learning techniques used for various disease prediction

	Machine learning technique	*Disease*	*Reference*
1	Artificial neural network (ANN), support vector machines (SVM) and decision tree	Cancer	[49]
2	ANN, SVM and random forest	Tuberculosis	[50]
3	Decision tree (DT), k-nearest neighbor (K-NN), multilayer perceptron (MLP), single conjunctive rule learner (SCRL), naïve Bayes (NB), radial basis function (RBF) and SVM	Heart disease	[51]
4	Principal component analysis (PCA) and minimum redundancy maximum relevance (mRMR)	Diabetes	[52]
5	Deep learning model	Skin cancer	[53]
6	SVM, logistic regression, ANN, K-NN, NB and decision tree	Heart disease	[54]
7	Decision trees, KNN, k-means and ANN	Thyroid	[55]
8	Classifiers (RBF, IBK and JRip)	Diabetes	[56]
9	Support vector machine	Breast cancer	[57]
10	SVM, ANN and logistic regression	Diabetes	[58]
11	Support vector machine	Coronary artery	[60]
12	Support vector machine	Alzheimer's	[60]
13	Random forest, linear regression (LR), multi-layer perceptron and decision trees	Breast cancer	[61]
14	Radial basis function (RBF), decision tree (DT), ANN, logistic regression (LR), Bayesian networks (BN) and SVM	Tuberculosis	[62]
15	Decision tree	Cancer	[63]

process. In another study, decision tree and naïve Bayes methods were used in the data mining technique [65] to retrieve the detailed information of every patient. Based on the precise consequence estimate, the performance of the classification was examined.

In the prediction of thyroid diseases, support vectors machines and decision trees classification algorithms were used by [66]. The results indicated 99.89% classification accuracy. Meanwhile, [67] discussed the method used to predict patients' thyroid disease, which was based on a number of thyroid symptoms. In this comparative study, different ML methods were used to accomplish improved prediction accuracy. Among many models, the decision tree has been found to be the best, with 99.46% accuracy. In the research study, thyroid disease is classified into three different groups: (i) hyperthyroidism, (ii) hypothyroidism, and (iii) the general category. To classify and separate these groups, [68] used different ML models, such as random forest, multi-layer perceptron, logistic regression, SVM, naïve Bayes, decision tree, k-nearest neighbor, and linear discriminant analysis.

Moreover, [69] presented a breast cancer prediction study using six supervised machine learning algorithms (logistic regression, k-nearest neighbor, random forest, support vector machine deep learning, decision tree). In this study, the highest accuracy (98.24%) was achieved by deep learning using the model.

Machine learning models were used by [70] to predict diabetes (type 2) complications. Diabetes is a foremost metabolic condition that can badly disturb the intact body system. In [71], five different predictive models were used to categorize diabetic and non-diabetic patients, operated in the R software. For this purpose, they used an artificial neural network, kernel support vector machine, k-nearest neighbor, linear kernel support vector machine, radial basis function, and multifactor dimension reduction techniques.

The [72] study identified molecules active against mycobacterium tuberculosis using machine learning techniques. The results of this research paper show the potential of the XGBoost model for the best prediction performance.

9.7 CONCLUSION AND FUTURE SCOPE

In the previous years, machine learning (ML) has changed from an educational interest to a multi-billion-dollar business and a point of attention for our scientific, socio-economic, and safety setup. Healthcare is one of the wildest rising divisions in the global economy; many people need care, and it is becoming gradually expensive. Government expenditure on healthcare has increased to an all-time high, while the underlying necessity to increase patient-doctor association becomes abundantly clear. This study displays that machine learning has given medical sources new tools to work with and new customs of medical practice. It also recognizes that ML tools and procedures are essential in healthcare provinces and are particularly used in the analysis and prediction of a wide variety of diseases. Advanced expertise, like big data and machine learning, have the prospective to equally support providers and patients in terms of better care at a lesser cost. As a result, effective machine learning methods in the healthcare industry become essential to address these trials. Machine learning and analytical methods can modernize the intact healthcare business by allowing precise vision and predictions associated with diagnosis, indicators, techniques, and treatments. In the future, many machine learning procedures can be used with better accuracy in prediction. Many corporations and organizations had previously taken the primary steps in this area and supported the changeover to patient and related attention. The future of ML in the healthcare sector is still undergoing excessive research. It can bring relief to medical ailments to a great extent. It could be very economical and efficient in the future. Various research are underway to use ML for cancer treatment, diabetes, heart disease, and others.

REFERENCES

[1] Bhardwaj, R., Nambiar, A. R., & Dutta, D. (2017, July). A study of machine learning in healthcare. In *2017 IEEE 41st Annual Computer Software and Applications Conference (COMPSAC)* (Vol. 2, pp. 236–241). IEEE.

[2] Gochhait, S., Singh, T., Bende, A., Thapliyal, M., Vemulapalli, H., Shukla, G., . . . & Chinta, D. (2020, June). Implementation of EHR using digital transformation: a study on telemedicine. In *2020 International Conference for Emerging Technology (INCET)* (pp. 1–4). IEEE.

[3] Fedushko, S., Ustyianovych, T., & Gregus, M. (2020). Real-time high-load infrastructure transaction status output prediction using operational intelligence and big data technologies. *Electronics*, 9(4), 668.

[4] Nithya, B., & Ilango, V. (2017, June). Predictive analytics in health care using machine learning tools and techniques. In *2017 International Conference on Intelligent Computing and Control Systems (ICICCS)* (pp. 492–499). IEEE.

[5] Raghupathi, W., & Raghupathi, V. (2014). Big data analytics in healthcare: promise and potential. *Health Information Science and Systems*, 2(1), 1–10.

[6] Pagliari, C. (2007). Design and evaluation in eHealth: challenges and implications for an interdisciplinary field. *Journal of Medical Internet Research*, 9(2), e614.

[7] Kannampallil, T. G., Franklin, A., Mishra, R., Almoosa, K. F., Cohen, T., & Patel, V. L. (2013). Understanding the nature of information seeking behavior in critical care: implications for the design of health information technology. *Artificial Intelligence in Medicine*, 57(1), 21–29.

[8] Conway, J., Johnson, B., Edgman-Levitan, S., Schlucter, J., Ford, D., Sodomka, P., & Simmons, L. (2006). *Partnering with patients and families to design a patient-and family-centered health care system: a roadmap for the future: a work in progress*. Institute for Family-Centered Care.

[9] Aceto, G., Persico, V., & Pescapé, A. (2018). The role of information and communication technologies in healthcare: taxonomies, perspectives, and challenges. *Journal of Network and Computer Applications*, 107, 125–154.

[10] Manteghinejad, A., & Javanmard, S. H. (2021). Challenges and opportunities of digital health in a post-COVID19 world. *Journal of Research in Medical Sciences: The Official Journal of Isfahan University of Medical Sciences*, 26.

[11] Holzinger, A. (2016). Interactive machine learning for health informatics: when do we need the human-in-the-loop?. *Brain Informatics*, 3(2), 119–131.

[12] Dunjko, V., & Briegel, H. J. (2018). Machine learning & artificial intelligence in the quantum domain: a review of recent progress. *Reports on Progress in Physics*, 81(7), 074001.

[13] Awad, M., & Khanna, R. (2015). *Efficient learning machines: theories, concepts, and applications for engineers and system designers* (p. 268). Springer Nature.

[14] Choy, G., Khalilzadeh, O., Michalski, M., Do, S., Samir, A. E., Pianykh, O. S., . . . & Dreyer, K. J. (2018). Current applications and future impact of machine learning in radiology. *Radiology*, 288(2), 318–328.

[15] Amrollahi, M., Hadayeghparast, S., Karimipour, H., Derakhshan, F., & Srivastava, G. (2020). Enhancing network security via machine learning: opportunities and challenges. *Handbook of big data privacy*, 165–189.

[16] Chinnamgari, S. K. (2019). *R Machine Learning Projects: Implement supervised, unsupervised, and reinforcement learning techniques using R 3.5*. Packt Publishing Ltd.
[17] Delmastro, F., Di Martino, F., & Dolciotti, C. (2020). Cognitive training and stress detection in mci frail older people through wearable sensors and machine learning. *IEEE Access*, 8, 65573–65590.
[18] Sathya, R., & Abraham, A. (2013). Comparison of supervised and unsupervised learning algorithms for pattern classification. *International Journal of Advanced Research in Artificial Intelligence*, 2(2), 34–38.
[19] Wang, Y., Martonosi, M., & Peh, L. S. (2007). Predicting link quality using supervised learning in wireless sensor networks. *ACM SIGMOBILE Mobile Computing and Communications Review*, 11(3), 71–83.
[20] Locatello, F., Bauer, S., Lucic, M., Raetsch, G., Gelly, S., Schölkopf, B., & Bachem, O. (2019, May). Challenging common assumptions in the unsupervised learning of disentangled representations. In *International conference on machine learning* (pp. 4114–4124). PMLR.
[21] Ramezani Dooraki, A., & Lee, D. J. (2018). An end-to-end deep reinforcement learning-based intelligent agent capable of autonomous exploration in unknown environments. *Sensors*, 18(10), 3575.
[22] Vithayathil Varghese, N., & Mahmoud, Q. H. (2020). A survey of multi-task deep reinforcement learning. *Electronics*, 9(9), 1363.
[23] Akanksha, E., Sharma, N., & Gulati, K. (2021, April). Review on reinforcement learning, research evolution and scope of application. In *2021 5th international conference on computing methodologies and communication (ICCMC)* (pp. 1416–1423). IEEE.
[24] Chugh, G., Kumar, S., & Singh, N. (2021). Survey on machine learning and deep learning applications in breast cancer diagnosis. *Cognitive Computation*, 13(6), 1451–1470.
[25] De, S., Das, R., Bhattacharyya, S., & Maulik, U. (Eds.). (2022). *Applied smart health care informatics: a computational intelligence perspective*. John Wiley & Sons.
[26] Hemanth, D. J. D. (Ed.). (2021). *Handbook of decision support systems for neurological disorders*. Academic Press.
[27] Bhardwaj, R., Nambiar, A. R., & Dutta, D. (2017, July). A study of machine learning in healthcare. In *2017 IEEE 41st annual computer software and applications conference (COMPSAC)* (Vol. 2, pp. 236–241). IEEE.
[28] Ngiam, K. Y., & Khor, W. (2019). Big data and machine learning algorithms for health-care delivery. *The Lancet Oncology*, 20(5), e262–e273.
[29] Siddique, S., & Chow, J. C. (2021). Machine learning in healthcare communication. *Encyclopedia*, 1(1), 220–239.
[30] Bhardwaj, R., Nambiar, A. R., & Dutta, D. (2017, July). A study of machine learning in healthcare. In *2017 IEEE 41st annual computer software and applications conference (COMPSAC)* (Vol. 2, pp. 236–241). IEEE.
[31] Degoulet, P., & Fieschi, M. (2012). *Introduction to clinical informatics*. Springer Science & Business Media.
[32] Naït-Ali, A., & Cavaro-Ménard, C. (Eds.). (2013). *Compression of biomedical images and signals*. John Wiley & Sons.

[33] Hsieh, M. H. (2010). *Evaluating the effect of a patient education brochure on patients' expectations and satisfaction with emergency department service* (Doctoral dissertation, Queensland University of Technology).
[34] Deo, R. C. (2015). Machine learning in medicine. *Circulation*, 132(20), 1920–1930.
[35] Jayatilake, S. M. D. A. C., & Ganegoda, G. U. (2021). Involvement of machine learning tools in healthcare decision making. *Journal of Healthcare Engineering*, 2021.
[36] Cruz, J. A., & Wishart, D. S. (2006). Applications of machine learning in cancer prediction and prognosis. *Cancer Informatics*, 2, 117693510600200030.
[37] Ahmed, Z., Mohamed, K., Zeeshan, S., & Dong, X. (2020). Artificial intelligence with multi-functional machine learning platform development for better healthcare and precision medicine. *Database*, 2020.
[38] Dunn, E. J., Mills, P. D., Neily, J., Crittenden, M. D., Carmack, A. L., & Bagian, J. P. (2007). Medical team training: applying crew resource management in the Veterans Health Administration. *The Joint Commission Journal on Quality and Patient Safety*, 33(6), 317–325.
[39] Liu, Y., Gao, J., Liu, J., Walline, J. H., Liu, X., Zhang, T., . . . & Zhu, W. (2021). Development and validation of a practical machine-learning triage algorithm for the detection of patients in need of critical care in the emergency department. *Scientific Reports*, 11(1), 1–9.
[40] Montori, V. (2020). *Why we revolt: a patient revolution for careful and kind care*. Rosetta Books.
[41] Fleisch, E., Franz, C., & Herrmann, A. (2021). *The digital Pill: what everyone should know about the future of our healthcare system*. Emerald Group Publishing.
[42] Tanwar, R., Balamurugan, S., Saini, R. K., Bharti, V., & Chithaluru, P. (Eds.). (2022). *Advanced healthcare systems: empowering physicians with IoT-enabled technologies*. John Wiley & Sons.
[43] Marwaha, J. S., Landman, A. B., Brat, G. A., Dunn, T., & Gordon, W. J. (2022). Deploying digital health tools within large, complex health systems: key considerations for adoption and implementation. *NPJ Digital Medicine*, 5(1), 1–7.
[44] Santosh, K. C., & Gaur, L. (2022). *Artificial intelligence and machine learning in public healthcare: Opportunities and societal impact*. Springer Nature.
[45] Beaulieu-Jones, B. K., Yuan, W., Brat, G. A., Beam, A. L., Weber, G., Ruffin, M., & Kohane, I. S. (2021). Machine learning for patient risk stratification: standing on, or looking over, the shoulders of clinicians?. *NPJ Digital Medicine*, 4(1), 1–6.
[46] Sutton, R. T., Pincock, D., Baumgart, D. C., Sadowski, D. C., Fedorak, R. N., & Kroeker, K. I. (2020). An overview of clinical decision support systems: benefits, risks, and strategies for success. *NPJ Digital Medicine*, 3(1), 1–10.
[47] Orlando, L. A., Buchanan, A. H., Hahn, S. E., Christianson, C. A., Powell, K. P., Skinner, C. S., . . . & Henrich, V. C. (2013). Development and validation of a primary care-based family health history and decision support program (MeTree). *North Carolina Medical Journal*, 74(4), 287–296.
[48] Piri, S., Delen, D., Liu, T., & Zolbanin, H. M. (2017). A data analytics approach to building a clinical decision support system for diabetic retinopathy: developing and deploying a model ensemble. *Decision Support Systems*, 101, 12–27.

[49] Shaikh, F. J., & Rao, D. S. (2022). Prediction of cancer disease using machine learning approach. *Materials Today: Proceedings*, 50, 40–47.
[50] Lai, N. H., Shen, W. C., Lee, C. N., Chang, J. C., Hsu, M. C., Kuo, L. N., . . . & Chen, H. Y. (2020). Comparison of the predictive outcomes for anti-tuberculosis drug-induced hepatotoxicity by different machine learning techniques. *Computer Methods and Programs in Biomedicine*, 188, 105307.
[51] Pouriyeh, S., Vahid, S., Sannino, G., De Pietro, G., Arabnia, H., & Gutierrez, J. (2017, July). A comprehensive investigation and comparison of machine learning techniques in the domain of heart disease. In *2017 IEEE Symposium on Computers and Communications (ISCC)* (pp. 204–207). IEEE.
[52] Zou, Q., Qu, K., Luo, Y., Yin, D., Ju, Y., & Tang, H. (2018). Predicting diabetes mellitus with machine learning techniques. *Frontiers in Genetics*, 515.
[53] Mijwil, M. M. (2021). Skin cancer disease images classification using deep learning solutions. *Multimedia Tools and Applications*, 80(17), 26255–26271.
[54] Li, J. P., Haq, A. U., Din, S. U., Khan, J., Khan, A., & Saboor, A. (2020). Heart disease identification method using machine learning classification in e-healthcare. *IEEE Access*, 8, 107562–107582.
[55] Razia, S., Siva Kumar, P., & Rao, A. S. (2020). *Machine learning techniques for Thyroid disease diagnosis: A systematic review*. Modern Approaches in Machine Learning and Cognitive Science: A Walkthrough, 203–212.
[56] Mishra, S., Chaudhury, P., Mishra, B. K., & Tripathy, H. K. (2016, March). An implementation of feature ranking using machine learning techniques for diabetes disease prediction. In *Proceedings of the second international conference on information and communication technology for competitive strategies* (pp. 1–3). https://doi.org/10.1145/2905055.2905100.
[57] Bayrak, E. A., Kırcı, P., & Ensari, T. (2019, April). Comparison of machine learning methods for breast cancer diagnosis. In *2019 Scientific meeting on electrical-electronics & biomedical engineering and computer science (EBBT)* (pp. 1–3). IEEE.
[58] Joshi, T. N., & Chawan, P. P. M. (2018). Diabetes prediction using machine learning techniques. *IJERA*, 8(1), 9–13.
[59] Abdar, M., Książek, W., Acharya, U. R., Tan, R. S., Makarenkov, V., & Pławiak, P. (2019). A new machine learning technique for an accurate diagnosis of coronary artery disease. *Computer Methods and Programs in Biomedicine*, 179, 104992.
[60] Trambaiolli, L. R., Lorena, A. C., Fraga, F. J., Kanda, P. A., Anghinah, R., & Nitrini, R. (2011). Improving Alzheimer's disease diagnosis with machine learning techniques. *Clinical EEG and Neuroscience*, 42(3), 160–165.
[61] Gupta, M., & Gupta, B. (2018, February). A comparative study of breast cancer diagnosis using supervised machine learning techniques. In *2018 second international conference on computing methodologies and communication (ICCMC)* (pp. 997–1002). IEEE.
[62] Kalhori, S. R. N., & Zeng, X. J. (2013). *Evaluation and comparison of different machine learning methods to predict outcome of tuberculosis treatment course* (pp. 1–10), https://doi.org/10.4236/jilsa.2013.53020.
[63] Sanyal, H., Saxena, P., & Agrawal, R. (2021). Innovative approach for prediction of cancer disease by improving conventional machine learning classifier. In *Cybernetics, Cognition and Machine Learning Applications* (pp. 281–289). Springer.

[64] Rani, K. U. (2011). Analysis of heart diseases dataset using neural network approach. *arXiv preprint arXiv:1110.2626*.
[65] Maheswari, S., & Pitchai, R. (2019). Heart disease prediction system using decision tree and naive Bayes algorithm. *Current Medical Imaging*, 15(8), 712–717.
[66] Dharmarajan, K., Balasree, K., Arunachalam, A. S., & Abirmai, K. (2020). Thyroid disease classification using decision tree and SVM. *Indian Journal of Public Health Research & Development*, 11(3), 229–233.
[67] Raisinghani, S., Shamdasani, R., Motwani, M., Bahreja, A., & Raghavan Nair Lalitha, P. (2019, April). Thyroid prediction using machine learning techniques. In *International Conference on Advances in Computing and Data Sciences* (pp. 140–150). Springer.
[68] Sonuç, E. (2021, July). Thyroid Disease Classification Using Machine Learning Algorithms. *Journal of Physics: Conference Series*, 1963(1), 012140. IOP Publishing.
[69] Gupta, P., & Garg, S. (2020). Breast cancer prediction using varying parameters of machine learning models. *Procedia Computer Science*, 171, 593–601.
[70] Dagliati, A., Marini, S., Sacchi, L., Cogni, G., Teliti, M., Tibollo, V., . . . & Bellazzi, R. (2018). Machine learning methods to predict diabetes complications. *Journal of Diabetes Science and Technology*, 12(2), 295–302.
[71] Kaur, H., & Kumari, V. (2020). Predictive modelling and analytics for diabetes using a machine learning approach. *Applied Computing and Informatics*. https://www.emerald.com/insight/content/doi/10.1016/j.aci.2018.12.004/full/html.
[72] Ye, Q., Chai, X., Jiang, D., Yang, L., Shen, C., Zhang, X., . . . & Hou, T. (2021). Identification of active molecules against Mycobacterium tuberculosis through machine learning. *Briefings in Bioinformatics*, 22(5), bbab068.

Chapter 10

Autism spectrum disorder diagnostic system using adaptive neuro fuzzy inference system

Joy Karan Singh and Deepti Kakkar

CONTENTS

10.1 Introduction 167
10.1.1 Autism spectrum disorder and its prevalence 167
10.1.2 Prevalence of autism and its subtypes 168
10.1.3 Autism diagnosis techniques 169
10.1.3.1 Primary diagnostic techniques 169
10.1.3.2 Fuzzy-based techniques for ASD diagnosis 170
10.2 Proposed diagnostic model 172
10.2.1 Implementation details 173
10.2.2 Input module 173
10.2.3 Fuzzifier module 174
10.2.4 Inference module 174
10.2.5 Defuzzification module 174
10.2.6 Decision module 174
10.2.7 Secondary module design 174
10.3 Results and discussion 175
10.4 Conclusion 177
Appendix A 177
References 178

10.1 INTRODUCTION

10.1.1 Autism spectrum disorder and its prevalence

Autism spectrum disorder is a complex disorder in which persistent challenges are introduced in communication, social interaction and repetitive patterns [1]–[2]. ASD can be defined as a triad of disorders: autism, Asperger's syndrome and pervasive developmental disorder

DOI: 10.1201/9781003305347-10

(PDD-NOS) [3]. The symptoms that are present in this disorder include the following:

1. Restrictive and repetitive pattern of behaviors, activities and interests [4].
2. Communication deficits which extend to verbal signals, such as development of speech signals. These deficits also extend to non-verbal signals, such as fixation of eyes and problems in facial expressions [7].
3. Body spinning, hand movements and rocking.
4. Eye contact avoidance and highly sensitive response to loud noises, toys and other changes for which normal individuals are comfortable [8].
5. Unusual reaction to changes in routine.

10.1.2 Prevalence of autism and its subtypes

The first instance of autism was first reported by a child psychiatrist in 1943, and then it was reported in 1944. The children reported in 1940 were 11 (8 boys and 3 girls). There has been a sudden increase in the number of cases of autism with 1 to 5 cases in 10,000 in the year 1972 [9]. Autism is more prevalent in boys compared to girls. The ratio of boys affected with autism is 2:1, compared to girls, in 1972. In the year 2006, it was reported that the ratio of boys affected were 3–4 times more than girls, with the ratio 3–4:1 (Table 10.1). The symptoms of autism being reported in females were mental retardation, along with impairments in social interaction, whereas genetic factors are the major symptoms in the case of males [9].

The variety in the criterion for the diagnosis of autism increased awareness among practitioners and families, and the accuracy of detection methods and professionals have increased the number of children affected by this autism [10]. This has sharply increased the number of autistic children, and per the Centers for Disease Control and Prevention (CDC), the number is 1 affected in 50 children in the U.S. per the latest record [11]. The following pie chart shows the proportion of different types of disabilities along with

Table 10.1 Prevalence of autism over the years

Surveillance year	*Birth year*	*Ratio of children affected*
2000	1992	1 in 150
2002	1994	1 in 150
2004	1996	1 in 125
2006	1998	1 in 110
2008	2000	1 in 88
2010	2002	1 in 68
2012	2004	1 in 69

their percentage disability. The percentage disability in seeing is 18.8, hearing is 18.9, speech is 7.5, movement is 20.3, mental retardation is 5.6, mental illness is 2.7, any other which includes autism also is 18.4 and multiple disabilities is 7.9 [12].

The subgroups of autism are discussed in Table 10.2.

10.1.3 Autism diagnosis techniques

10.1.3.1 Primary diagnostic techniques

Primarily, ASD can be diagnosed using traditional screening tools such as scales, interviews and surveys [12]. Among these methods, scales are the widely preferred method by psychologists and clinicians which include the Child Behavior Checklist (CBCL for age group 6–18), Vanderbilt's Assessment Scale (for age group 6–12), Behavior Assessment System (BASC-3 for age group 2–21), Connor's Comprehensive Behavior Rating Scale (CBRS for age group 6–18) [13]. These scales are based on a grading system, wherein the severity levels of each symptom are manually quantified using grades and a cumulative score. Each scale has its predefined threshold value which are compared with the cumulative score. Based on the comparison with threshold values, the child under investigation is categorized as normal, medium or severe. These scales yielded moderate sensitivity (83.0%) and specificity (84.0%) in ASD diagnosis [14], and sometimes, the analysis of multiple minor deficits can result in the same scoring, and results can vary in case of improper knowledge of parents or guardians of these scales [15]. Thus, the screening scales rely primarily on the series of subjective evaluations and highly depends on the doctor's expertise and intuitions; due to which, several non-invasive brain imaging methods, such as EEG/ERP, functional magnetic resonance (fMRI), positron emission tomography (PET) [16] and others, are being exploited to study and understand the neural markers of ASD.

Among the imaging methods, EEG provides a high temporal resolution and potential in diagnosing ASD. For evaluating the EEG rhythms, a large number of electrodes are required which could be uncomfortable for ASD

Table 10.2 Percentage value of subgroups of autism

Subgroup of autism	*Percentage (%)*	*Source*
ASD with genetic disorder	20	(Sinha et al., 2014) [5]
ASD with anxiety	84	(Song, Liu and Kong, 2016) [7]
ASD with aggressive/ danger-free behavior	38	(Dominick et al., 2007) [8]
ASD with epilepsy	42	(Song, Liu and Kong, 2016) [10]

Note: The major portion of the disorder covered by co-morbidities and the review of studying these behavioral traits leading to disorder diagnosis have been provided in the literature survey.

children at the social and personal levels, so researchers are focusing on ERP which requires a few numbers of electrodes focusing only on specific regions of abnormalities. Hence, the present paper is focusing on auditory ERP since the number of electrodes is minimal and analysis can be easily interpreted. An ERP is an evaluation of a brain response in terms of potential that is the direct result obtained through a specific cognitive, sensory or motor event [17]. It is a sophisticated method for extracting task-related specific sensory, cognitive and motor response [18] by using simple averaging techniques. The ERP signals were recorded, and features that indicate their attention and cognitive abilities, such as the P300 wave [19], are extracted to discriminate between normal and ASD groups. P300 latency is considered as a psycho-physiological signature [20] to determine deficits showing inhibitory control. The reduced P300 peak amplitude elicited in auditory oddball tasks show a significant difference between the ASD and normal group. A small or missing P300 wave reflects impairments in behavioral control.

The different primary and secondary diagnostic techniques have been discussed in tabular form in Table 10.3 and Table 10.4, respectively. The table data also provides the validity, reliability and diagnostic criteria followed by the scales [20]. This section describes the different diagnostic methods for detection of neurodevelopmental disorders (ASD).

(AS: Asperger's syndrome, PDD-NOS: pervasive developmental disorder not otherwise specified, ID: intellectual disability, ADHD: attention deficit hyperactive disorder)

10.1.3.2 Fuzzy-based techniques for ASD diagnosis

In this next section, we propose a fuzzy-based model for the detection of autism. The fuzzy logic model can be viewed as a way to convert expert knowledge into an automatic control strategy without any detailed knowledge of the problem. It is mainly composed of four components (fuzzification, defuzzification, database, fuzzy rule–base inference engine), as shown in Figure 10.3. In this section, we proposed that the rule-based fuzzy diagnostics is a decision support system that intended for physicians; this will analyse the stage of ASD. Fuzzy systems are fuzzy model structures in the form of fuzzy rule bases (FRBs) that are the most important area in the application of the fuzzy set theory [21]. Designing a fuzzy rule–based system involves deriving the desired if-then fuzzy rules, partitioning universes, and addressing membership functions. Fuzzy rule base is updated with dynamic rules. ASD symptoms and stage details are updated in the rule base. As the patient must have a history in ASD, then the proposed system uses fuzzy logic as the tool to develop the algorithm of the system.

Table 10.3 Primary diagnostic techniques for ASD

	Sample characteristics			*Type*	*Purpose*	*Advantages*	*Disadvantages*
	Age	*Size*	*Nature*				
David F. D. et al. [7]	18–30	10 adults	ASD		Microstate analysis was carried out.		
Matson et al., 2007 [29]	20–80 years	232 adults	ID, autism and PDD-NOS	Rating scale (31 items)	To differentiate between individuals with ID having ASD and not having ASD.	Optimally differentiated the patients with ASD from those without ASD.	No accurate differentiation between autism and PDD-NOS on the basis of communication deficit.
Allison et al. [18]	18–24 months	160 children	Autism spectrum condition (ASC)	Rating scale in form of a questionnaire (25 items)	To identify toddlers at risk of autism with higher sensitivity in clinical as well as research motive.	First screening instrument which demonstrated deficits of autism at 18 months of age and for dimensioning traits of autism.	Cannot be used for clinically screening the disorders.
Sappok et al. [15]	35 years (mean)	219 adults	ID	Scale (20 items)	To assess features of autism in individuals with ID.	Useful and psychometrical instrument for screening ASD in adults with ID.	Screening instrument, not a diagnostic instrument.

Table 10.4 Secondary diagnostic techniques for ASD

Reference	*Experiment*	*Participants*	*Parameters studied*	*Deficit found*
Falck-Ytter et al., 2010 [6]	Video with object-directed actions	18 ASD Mean age: 5.1 years	Eye movements and role of repetition	Intact mirror neuron system
Vivanti et al., 2011 [24]	Four-second video of goal-directed actions	18 ASD Mean age: 13 years	Eye movement and behavioral measures	Poor social interpretation
Marsh et al., 2011 [8]	Movie clips depicting two triangles with different interaction styles	18 ASD Mean age: 33 years	Reaction time, repetition effect	Mirror neuron system is affected
Hofsten et al., 2012 [22]	Conversational session	9 ASD Mean age: 3–6 years old	Eye fixations and gaze patterns	Social interactional functioning and prospective control deficit
Palmer et al., 2015 [20]	Hypotheses: extent of awareness and understanding of other's mind states	–	–	Social cognition deficit due to reduced understanding of other's mind states
Schuwerk et al., 2016 [9]	Introductory movies	26 ASD Mean age: 9.7 years	Eye movements	Social interaction deficit
Chambon et al., 2017 [10]	Social and non-social intention tasks	18 ASD Mean age: 35.7 years	Discriminability index (d0) response bias and learning effect	Restricted and repetitive behavior

10.2 PROPOSED DIAGNOSTIC MODEL

The proposed diagnostic model is divided into two stages, viz. primary screening stage and secondary diagnostic stage. The primary stage predicts ASD based on three behavioral ASD responses, viz. inattention, hyperactivity/impulsivity and oppositional factors. A questionnaire is prepared based on the well-known Connor's scale (short version) for ASD diagnosis, as described in Appendix A. Three correlated functioning fuzzy inferences, viz. inattention, hyperactivity/impulsivity and oppositional aspects module are created with their specified expertise ranges in the form of trapezoidal membership functions. The three categories were then implemented into hierarchical fuzzy systems to reduce the number of rules and ease in generalizing results, as shown in Figure 10.1. The diagnostic procedure is carried to the secondary stage if the primary stage result lies either in the medium or high severity level.

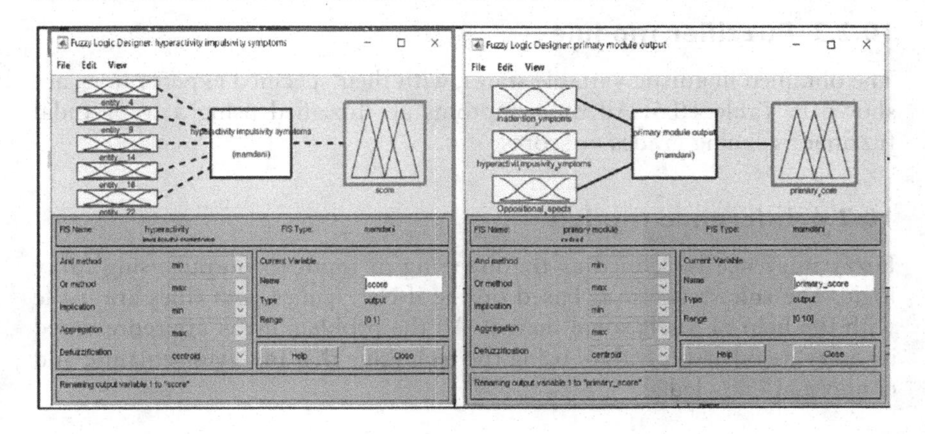

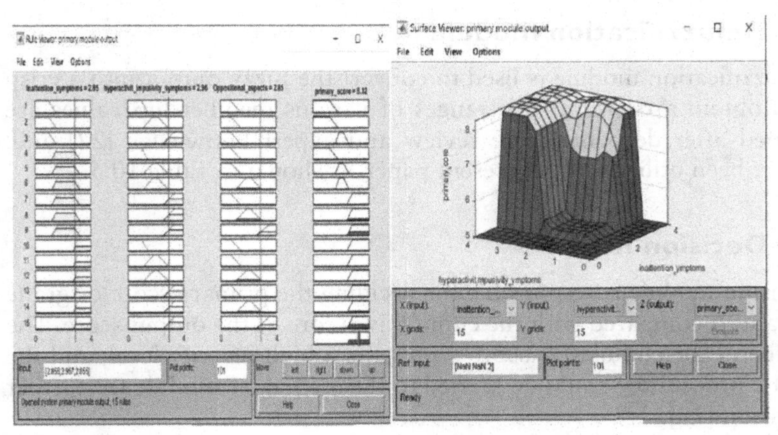

Figure 10.1 (a) FIS for hyperactivity/impulsivity module; (b) primary module with its three inputs, viz. inattention symptoms, hyperactivity/impulsivity and oppositional aspects.

10.2.1 Implementation details

The primary module consists of five modules, viz. input module, fuzzifier module, inference module, defuzzification and decision module. This section shows the function of each module for the primary screening.

10.2.2 Input module

This unit is responsible for creating the input vector matrix for our fuzzifier. The range of the linguistic variable is from 0 to 4. The parent or teacher is required to give a score based on the questionnaire provided, and this score is utilized as the input to the fuzzifier module.

10.2.3 Fuzzifier module

The obtained linguistic variable states, with their specified expert range, are shown in Table 10.5. All the symptoms are fuzzified using a trapezoidal fuzzifier assigning grades or scores.

10.2.4 Inference module

Fuzzy inference is defined as the mapping of input to output using fuzzy logic. The rule structure is based on the if-else syntax, and rules are made with the help of a relational matrix. All the problem areas are represented as fuzzy membership values using the technique that involves semantic and contextual knowledge.

10.2.5 Defuzzification module

The defuzzification module is used to convert the fuzzy output into a crisp value to obtain a decision. The ranges of various membership values are determined after deep literature review and expert knowledge [22] and, thus, have been utilized in the present paper, as shown in Table 10.5.

10.2.6 Decision module

The decision module is responsible for deciding the primary score or grade of ASD and is compared with the cumulative score of the output score. The high value of the cumulative score represents a high severity level, and the diagnostic procedure should be carried to the secondary module for further investigation [23].

10.2.7 Secondary module design

The secondary stage is designed to train the system with the ANFIS classifier utilizing the data obtained from the ERP analysis. There are two input variables, viz. peak amplitude and latency rate of P300 wave. These input variables are fed into the input layer, and the system is trained with

Table 10.5 Linguistic variable state with their specified ranges for determining primary stage output

	Behavioral variables	*Range*
1	Low	0–1.5
2	Medium	1.20–3.12
3	High	2.57–4.0

a backpropagation algorithm to obtain the minimum error. Table 10.6 is a cumulative of the specified ranges of the latency rate and peak amplitude of P300 for determining ASD severity level for the secondary stage.

10.3 RESULTS AND DISCUSSION

The result section has been analysed in two stages, viz. primary stage and secondary stage. The first stage uses the hierarchical version of Connor's scale, and the secondary stage utilizes the ERP score based on the oddball paradigm [24],[25]. The overall result is the score obtained from the secondary module. The output of the three correlating primary fuzzy modules is provided as the input to the final primary module per the hierarchical design. The primary final output is then defuzzified to obtain the crisp value. Figure 10.2 (a, b) shows the rule viewer and surface viewer demonstrating the relationship of the input variable with its output of the primary stage [26]. The primary stage output predicts whether the children show a low,

Table 10.6 *Severity level range for secondary stage output determination*

	Severity level or linguistic states	*Latency rate range (ms)*	*Peak amplitude range (μV)*
1	Normal	300–350	11–18
2	Medium	330–380	5–12
3	High	375–450	0–6

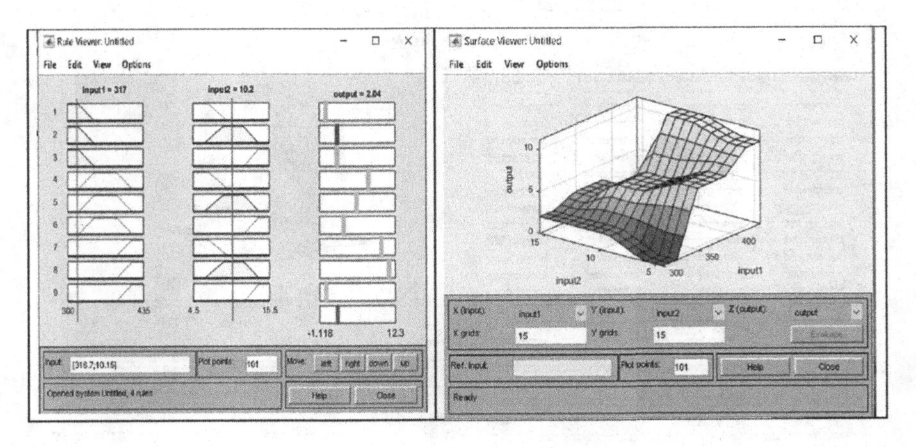

Figure 10.2 (a) Rule viewer for the primary decision, (b) surface viewer for the primary decision.

medium or high characteristic of ASD symptoms [27]. The secondary module evaluates the ERP scores for the final decision. Table 10.7 shows the average latency rate and peak amplitude of the P300 wave at the Fz, Pz and Cz electrode sites [28].

The ANOVA method confirms that there is a significant difference between ASD and normal children by evaluating the means and standard deviation ($p<0.01$) [29]. Hence, the statistical null hypothesis is rejected. Similar trends were observed on the Connor's scale but with less accuracy.

The system showed an accuracy of 99.3% in the classification (i.e., ASD vs. normal and) 88.78% in the severity level (normal/low, medium and high) of ASD [30]. This indicates that our proposed system differentiates between ASD and normal children with high accuracy [31].

Figure 10.3 (a) shows the rule viewer of the secondary module output. The output severity level ranges from 0 to 10, reflecting normal, medium and high severity levels. Figure 10.3 (b) showcases the relationship between

Table 10.7 Comparison of average latency rate and average amplitude in normal group vs. ASD group

Electrode site	*Average latency rate (ms)*		*Average amplitude (μV)*	
	Normal group	*ASD group*	*Normal group*	*ASD group*
Fz	330 ± 24.5	379.53 ± 27.9	15.7 ± 4	10.7 ± 3.75
Pz	332 ± 20.4	380.95 ± 35.5	15.3 ± 7.5	9.65 ± 2.5
Cz	329.5 ± 30.5	383.75 ± 49.5	13.5 ± 2.5	10.2 ± 4.5

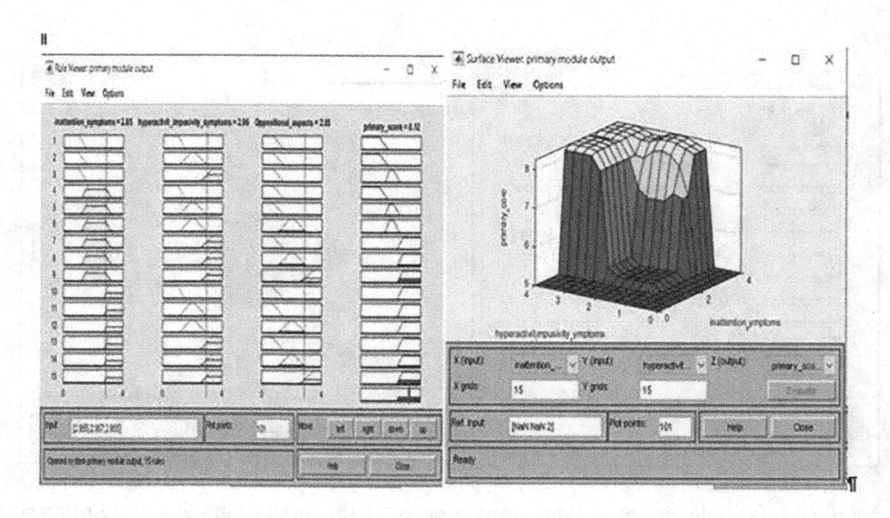

Figure 10.3 (a) Output rule viewer for the secondary module, (b) output surface viewer.

the output values and their corresponding input values. The decision can be made based on these obtained values, and according to their severity level, the children can be classified into different groups.

10.4 CONCLUSION

The proposed diagnostic module examines both the behavioral aspects as well as the brain's activity response under different psychiatric conditions (ERP). The primary modules make the decision and check the severity level of ASD. The procedure is carried out to the secondary module if the children show medium or high severity level. The secondary module evaluates the children's brain activity based on the well-known oddball paradigm and trains the artificial network, which makes decision-making easier. The secondary module also classified the children according to their severity level from 1 to 10 in terms of their attention and perceptional abilities. The proposed diagnostic module is time-efficient and can predict ASD at an early stage. Thus, it can be used as an adaptive automatic tool to provide relevant diagnostic information without ambiguity or change in the opinions of experts.

APPENDIX A

Category	*Entity*	*Symptoms*
Hyperactivity/impulsivity	Entity_4	Hyperactivity
	Entity_9	Speech delay
	Entity_5	Anxiety
	Entity_18	Change in voice
	Entity_22	Self-harm
	Entity_26	Difficulty in playing or engaging in leisure activity quietly
Inattention	Entity_3	Has difficulty in completing homework
	Entity_8	Often fails to complete the assignment on time
	Entity_12	Needs parents' supervision in completing assignments
	Entity_17	Avoids engaging in activities requiring mental efforts
	Entity_21	Unable to concentrate in class
	Entity_25	Does not follow instructions in school or workplace
Oppositional aspects	Entity_4	Inappropriate social interaction
	Entity_9	Poor eye contact
	Entity_14	Impulsivity

REFERENCES

[1] M.S. Mythili and A.R. Mohamed, "An improved autism predictive mechanism among children using fuzzy cognitive map and feature extraction methods (FEAST)," *ARPN Journal of Engineering and Applied Sciences*, vol. 11, 2016, pp. 1819–1828.

[2] E. Grossi, C. Olivieri and M. Buscema, "Diagnosis of autism through EEG processed by advanced computational algorithms: A pilot study," *Comput Methods Progr Biomed*, vol. 142, 2017, pp. 73–79.

[3] W. Hirstein, P. Iversen and V.S. Ramachandran, "Autonomic responses of autistic children to people and objects," *Proceedings of the Royal Society of London*, Series B, vol. 268, no. 1479, 2001, pp. 1883–1888. https://doi.org/10.1098/rspb.2001.1724.

[4] W.K. Shams, A. Wahab and U.A. Qidwai, "Fuzzy model for detection and estimation of the degree of autism spectrum disorder," in *International Conference on Neural Information Processing*, Berlin, Heidelberg: Springer, 2012, November, pp. 372–379.

[5] M. Ahmadlou, H. Adeli and A. Adeli, "Fractality and a waveletchaos-neural network methodology for EEG-based diagnosis of autistic spectrum disorder," *Journal of Clinical Neurophysiology*, vol. 27, 2012, pp. 328–333.

[6] A. Pratap, C.S. Kanimozhiselvi, R. Vijayakumar and K.V. Pramod, "Parallel neural fuzzy-based joint classifier model for grading autistic disorder soft computing applications," *Advances in Intelligent Systems and Computing*, vol. 356, 2016, pp. 13–26. https://doi.org/10.1007/978-3-319-18296-4_2

[7] R.M. Joseph, K. Ehrman, R. Mcnall and B. Keehn, "Affective response to eye contact and face recognition ability in children with ASD," *Journal of the International Neuropsychological Society*, vol. 14, no. 6, 2008, pp. 947–955. https://doi.org/10.10170S1355617708081344

[8] N. Kanwisher and G. Yovel, "The fusiform face area: A cortical region specialized for the perception of faces," *Philosophical Transactions of the Royal Society of London, Series B, Biological Sciences*, vol. 361, no. 1476, 2006, pp. 2109–2128. https://doi.org/10.1098/rstb.2006.1934.

[9] A. Kylliäinen and J.K. Hietanen, "Skin conductance responses to another person's gaze in children with autism," *Journal of Autism and Developmental Disorders*, vol. 36, no. 4, 2006, pp. 517–525. https://doi.org/10.1007/s10803-006-0091-4.

[10] P.J. Lang, "Emotion and motivation: Attention, perception, and action," *Journal of Sport and Exercise Psychology*, vol. 22, no. S1, 2000, pp. 122–140.

[11] R.D. Pascual-Marqui, "Discrete, 3D distributed, linear imaging methods of electric neuronal activity," *Part 1: Exact, Zero Error Localization*, 2007, https://arxiv.org/abs/0710.3341

[12] M.J. Carter, "Diagnostic and statistical manual of mental disorders," *Therapeutic Recreation Journal*, vol. 48, no. 3, 2014, pp. 275–277.

[13] American Psychiatric Association, *Diagnostic and statistical manual of mental disorders: DSM-5* (5th ed.). Arlington, VA: American Psychiatric Association, 2013.

[14] A.M. Tharpe et al., "Auditory characteristics of children with autism," *Ear and Hearing*, vol. 27, no. 4, 2006, pp. 430–441.

[15] A. Tas, R. Yagiz, M. Tas, M. Esme, C. Uzun and A.R. Karasalihoglu, "Evaluation of hearing in children with autism by using TEOAE and ABR," *Autism*, vol. 11, no. 1, 2007, pp. 73–79.
[16] B. H. Brask, "A prevalence investigation of childhood psychoses," *Nordic Symposium on the Comprehensive Care of the Psychotic Children*, 1972, pp.145–153.
[17] B. Kadesjö et al., "Brief report: Autism and Asperger syndrome in seven-year-old children: A total population study," *Journal of Autism and Developmental Disorders*, vol. 29, no. 4, 1999, pp. 327–331.
[18] C. Allison, S. Baron-Cohen, S. Wheelwright, T. Charman, J. Richler, G. Pasco and C. Brayne, "The Q-CHAT (Quantitative Checklist for Autism in Toddlers): A normally distributed quantitative measure of autistic traits at 18–24 months of age: Preliminary report," *Journal of Autism and Developmental Disorders*, vol. 38, no. 8, 2008, pp. 1414–1425.
[19] C. Gillberg et al., "Outcome and Prognostic factors in infantile autism and similar conditions: A population based study of 46 cases followed through puberty," *Journal of Autism and Developmental Disorder*, vol. 17, no. 2, 1987, pp. 273–287.
[20] C.J. Palmer, A.K. Seth and J. Hohwy, "The felt presence of other minds: Predictive processing, counterfactual predictions, and mentalising in autism," *Consciousness and Cognition*, vol. 36, 2015, pp. 376–389.
[21] C. Gillberg et al., "Outcome and Prognostic factors in infantile autism and similar conditions: A population based study of 46 cases followed through puberty," *Journal of Autism and Developmental Disorder*, vol. 17, no. 2, 1987, pp. 273–287.
[22] C.J. Palmer, A.K. Seth and J. Hohwy, "The felt presence of other minds: Predictive processing, counterfactual predictions, and mentalising in autism," *Consciousness and Cognition*, vol. 36, 2015, pp. 376–389.
[23] C. Von Hofsten and K. Rosander, "Perception-action in children with ASD," *Frontiers in Integrative Neuroscience*, vol. 6, 2012.
[24] D.L. Robins, K. Casagrande, M. Barton, C. Chen, T. Dumont-Mathieu and D. Fein, "Validation of the modified checklist for autism in toddlers, revised with follow-up (M-CHAT-R/F)," *Pediatrics*, vol. 133, no. 1, 2014, pp. 37–45.
[25] G. Vivanti, C. McCormick, G.S. Young, F. Abucayan, N. Hatt, A. Nadig and S.J. Rogers, "Intact and impaired mechanisms of action understanding in autism," *Developmental Psychology*, vol. 47, no. 3, 2011, p. 841.
[26] G. Zhu, Y. Li and P.P. Wen, "Epileptic seizure detection in EEGs signals using a fast weighted horizontal visibility algorithm," *Computer Methods and Programs in Biomedicine*, vol. 115, no. 2, 2014, pp. 64–75.
[27] H. Elhabashy et al., "Quantitative EEG in autistic children," *The Egyptian Journal of Neurology, Psychiatry and Neurosurgery*, vol. 52, no. 3, pp. 176–182, April 2015.
[28] L. Kanner, "Autistic disturbances of affective contact," *Nervous Child*, vol. 2, 1943, pp. 217–250.
[29] J. Bertrand et al., "Prevalence of autism in a United States population: The Brick Township, New Jersey, investigation," *Pediatrics*, vol. 108, no. 5, pp. 1155–1161, 2001.
[30] J.L. Matson, M. Nebel-Schwalm and M.L. Matson, "A review of methodological issues in the differential diagnosis of autism spectrum disorders in children," *Research in Autism Spectrum Disorders*, vol. 1, no. 1, 2007, pp. 38–54.

[31] J.L. Matson and M. Shoemaker, "Intellectual disability and its relationship to autism spectrum disorders," *Research in Developmental Disabilities*, vol. 30, no. 6, 2009, pp. 1107–1114.
[32] K.C. Chua, V. Chandran, U.R. Acharya and C.M. Lim, "Application of higher order spectra to identify epileptic EEG," *Journal of Medical Systems*, vol. 35, no. 6, 2011, pp. 1563–1571.

Chapter 11

Detection of diabetic foot ulcer (DFU) with AlexNet and ResNet-101

Hassana Abubakar, Zubaida Sa'id Ameen, Sinem Alturjman, Auwalu Saleh Mubarak, and Fadi Al-Turjman

CONTENTS

11.1 Introduction 181
11.1.1 Related work 182
11.2 Methodology 182
11.2.1 Data retrieval and processing 182
11.2.2 Models 183
11.2.3 Model training 184
11.2.4 Performance evaluation 184
11.3 Results and discussion 184
11.4 Conclusion 189
References 189

11.1 INTRODUCTION

Diabetes mellitus is one of the leading causes of deaths worldwide [1]. Diabetic foot ulcer (DFU) is an open wound that could possibly grow on a diabetic patient's feet [2]. Changes in skin color, skin temperature variation, feet swelling, dry cracks and leg pains are the main symptoms of DFU [3]. Diabetic patients have a 15% to 25% chance of developing DFU which can lead to lower limb elimination if not properly handled [4]. Hence, the survival rates of these patients are definitely reduced [5].

Annually, over 1 million diabetic patients have been reported to have limb amputation as a result of inappropriate DFU recognition and treatment [6]. The detection of DFU is challenging because it appears in different locations, sizes and shapes, and its precise diagnosis needs complete medical history and tests which are not available in most developing countries due to high cost [3], [7]. Thus, computerized methods are needed to overcome these challenges for early identification of DFU. This study aimed at detecting DFU using AlexNet and ResNet models at different epoch and data split.

DOI: 10.1201/9781003305347-11

Machine learning is an artificial intelligence field which provides computers with the capability of learning without them being programmed explicitly [8]. Deep learning is a subset of machine learning which accepts and processes data like the human brain via neural network. Convolutional neural network is a technique which has made deep learning popular recently. It is a neural network first suggested for image classification which contains one or more convolutional layers and pooling layers. It is very important in the classification of medical images [9], [10].

There are various architectures of CNN with very similar basic components, normally made up of three layers: convolution, pooling and fully connected layers. The available best models are AlexNet, VGGNet, GoogleNet and ResNet, having 8 layers, 19 and 16 layers, 22 layers and 9 modules, and 152 layers, respectively [11], [12]. Artificial intelligence is applied in different disciplines to solve problems related to gene editing, Covid-19 classification, smart tourism and water treatment [13]–[17].

11.1.1 Related work

Due to the health risk of DFU, various models have been designed to aid in the early and easy detection of this condition. YOLOv2-DFU models have been developed by Amin et al. [3] to classify DFU. This model contains 16 layers of convolutional neural network. And [18] made a suggestion that improved performance is not obtained by making a traditional CNN very deep with more layers and, therefore, designed a new deep CNN DFU_QUT-Net with only an increased width for classification of healthy skin and DFU skin. The performance of this model was also compared with AlexNet and VGGNet. Also, [19] developed a new CNN architecture; the performance of the DFUNet model for the classification of DFU was compared with that of conventional machine learning methods, AlexNet and GoogleNet. Then [20] proposed a deep learning network to classify thermograms of a diabetic foot. Transfer learning approach in terms of AlexNet and GoogleNet was also tested in this study. Further, [21] designed another CNN DFU model by introducing convolutional layer parallel branches between 2 and 5 into the old-style convolutional layers in which four branch models gave a better result. Faster R-CNN, 19, EfficientDet and YOLOv5 have been employed by [22] to identify the standard of their new dataset. However, the studies so far have not combined AlexNet and ResNet, and compared their performance at different parameters in the classification of DFU which is the focus of this study.

11.2 METHODOLOGY

11.2.1 Data retrieval and processing

The foot image dataset used in this study was obtained from the Kaggle database. This dataset originally contains 73 images of DFUs and 61 healthy skin. Figure 11.1 shows a DFU image and a normal foot image. The images were augmented using RandXReflection, RandXTranslation, RandYTranslation and RandRotation. Our proposed approach in this study is transfer learning. Transfer learning is a machine learning technique through which a new task learning is improved by transferring previously learned related task knowledge [23]. Efficient performance has also been observed in transfer learning on deep learning when little dataset was used, unlike models constructed from scratch that needs a large dataset for best performance [24]. The images were resized as 227 × 227 × 3 for AlexNet input size [25] and 224 × 224 × 3 for ResNet-101 input size [26] to be used in this study.

11.2.2 Models

AlexNet is a deep neural network that uses rectified linear unit (ReLU) as an activation function. The model has 227 × 227 × 3 size as a standard input, and its 8 layers are composed of 5 convolutional and 3 fully connected layers, including an output layer (Figure 11.2). The 5 convolutional layers have 3 × 3 filter sizes without padding and a max pooling operation window size of 2 × 2. It uses the Softmax classifier in the output layer [25]. AlexNet has

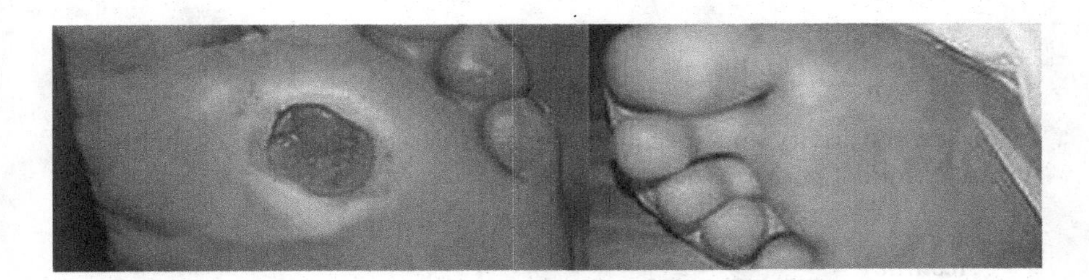

Figure 11.1 Left, diabetic foot ulcer; right, healthy skin.

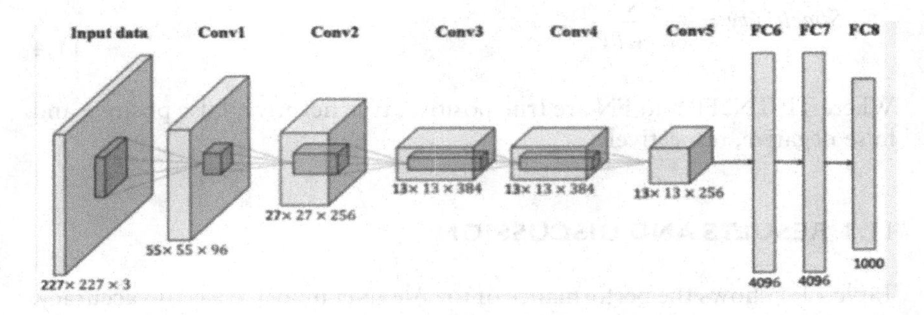

Figure 11.2 AlexNet architecture [33].

been successfully used in extensive image classification, including cancer [27], pneumonia [28], tuberculosis [12], chest X-rays [29] and skin lesions [30].

ResNet is made up of numerous fundamental residual blocks that give a shortcut connection between layers and does not permit the build-up of error on the convolution layers. The shortcut links allow the establishment of layers (hundred or more) and acquire greater performance in the process. This model is aimed at classifying extensive data. ResNet-101, ResNet-50, ResNet-32 and ResNet-18 are mostly employed for classification of medical images [31], [32].

11.2.3 Model training

MATLAB was used to train the model on a personal computer with 8 GB RAM and 64-bit Windows. To evaluate and obtain optimum performance of the model, the images were split into two different groups for training and testing: 70% training, 30% testing; and 80% training, 20% testing. Also, different epochs were employed with 32 minibatch size and 0.0001 initial learning rate.

11.2.4 Performance evaluation

To evaluate the loss of a trained model, the following equation is used:

$$\text{Loss} = \frac{-1}{n}\sum_{i=1}^{n} logPC \qquad 11.1$$

Accuracy, sensitivity and specificity are considered during the evaluation of a trained model performance:

$$Accuracy = \frac{TP + TN}{TP + TN + FP + FN} \qquad 11.2$$

$$Sensitivity = \frac{TP}{TP + FN} \qquad 11.3$$

$$Specificity = \frac{TN}{TN + FP} \qquad 11.4$$

Where TP, TN, FP and FN are true positive, true negative, false positive and false negative, respectively.

11.3 RESULTS AND DISCUSSION

Table 11.1 shows the performance of the AlexNet model, based on accuracy, sensitivity and specificity at 70:30 split, 80:20 split and 20 and 40 epochs. At 70:30 split, high performance was attained at 40 epoch, having 92.5%,

90.9% and 94.4% accuracy, sensitivity and specificity, respectively. At 80:20 split, the optimum performance was also attained at 40 epochs with higher accuracy, sensitivity and specificity of 97.1%, 100% and 91.7%, respectively. The performance of AlexNet with training and testing at 70:30, with training epochs 20 and 40, are shown in Figure 11.3 and Figure 11.4, respectively. While Figure 11.5 and Figure 11.6 show the performance of AlexNet with 80:20 data split at 20 and 40 epochs, respectively.

Table 11.1 AlexNet performance evaluation

Ratio	*Epoch*	*Accuracy*	*Sensitivity*	*Specificity*
70:30	20	87.5	86.4	88.9
	40	92.5	90.9	94.4
80:20	20	85	86.4	83.3
	40	97.1	100	91.7

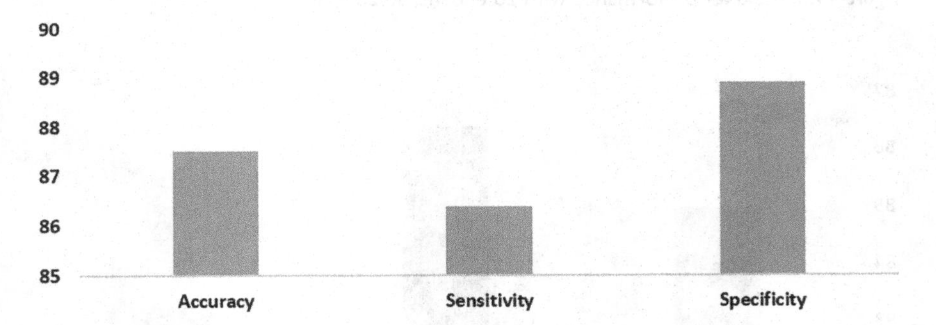

Figure 11.3 AlexNet performance with 20 epochs, 70:30 split.

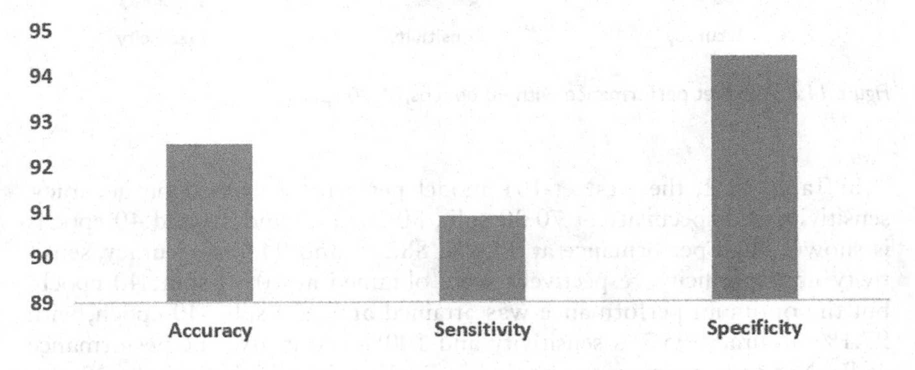

Figure 11.4 AlexNet performance with 40 epochs, 70:30 split.

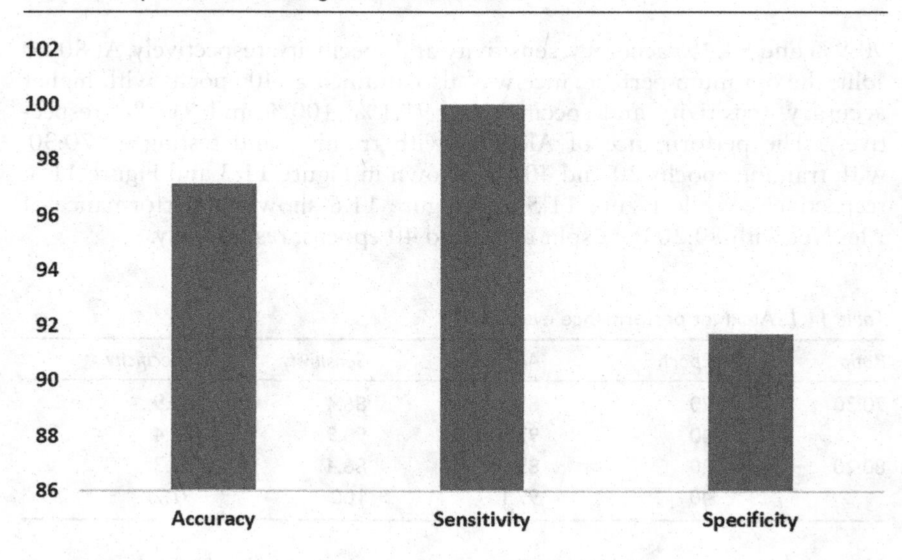

Figure 11.5 AlexNet performance with 20 epochs, 80:20 split.

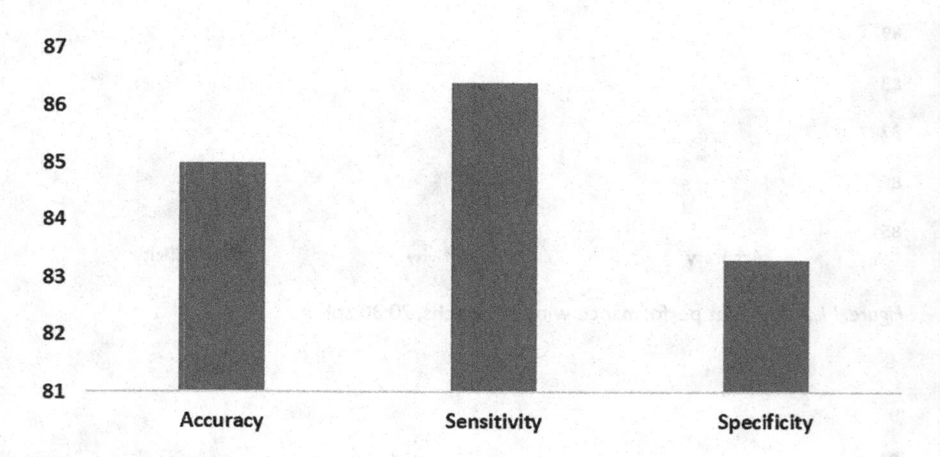

Figure 11.6 AlexNet performance with 40 epochs, 80:20 split.

In Table 11.2, the ResNet-101 model performance based on accuracy, sensitivity and specificity at 70:30 split, 80:20 split, and 20 and 40 epochs is shown. High performance at 90.4%, 88.2% and 94.4% accuracy, sensitivity and specificity, respectively, were obtained at 70:30 split, 40 epoch; but the optimum performance was attained at 80:20 split, 40 epoch, with 97.1% accuracy, 95.7% sensitivity and 100% specificity. The performance of ResNet with training and testing at 70:30, with training epochs 20 and 40, are shown in Figure 11.7 and Figure 11.8, respectively. While Figure 11.9

Table 11.2 ResNet performance evaluation

Ratio	*Epoch*	*Accuracy*	*Sensitivity*	*Specificity*
70:30	20	86.6	100	61.1
	40	90.4	88.2	94.4
80:20	20	85.7	87	83.3
	40	97.1	95.7	100

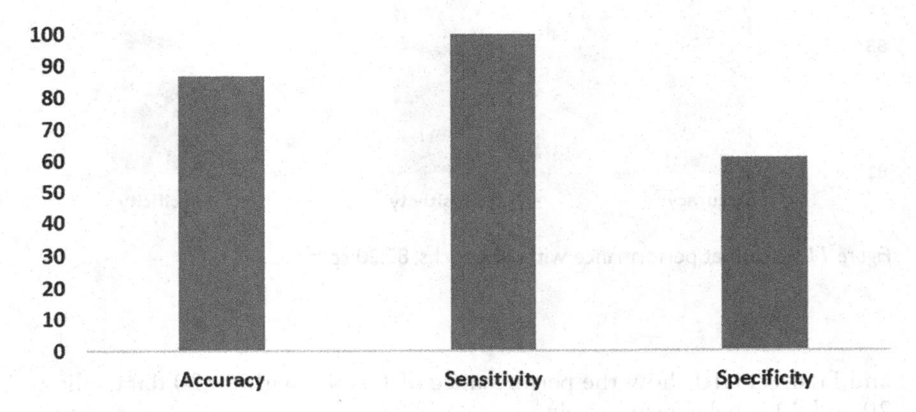

Figure 11.7 ResNet performance with 20 epochs, 70:30 split.

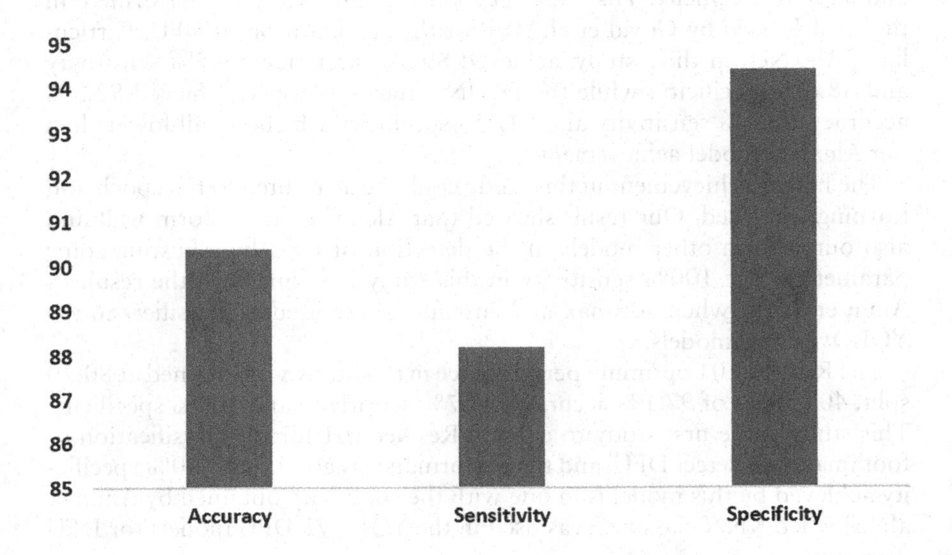

Figure 11.8 ResNet performance with 40 epochs, 70:30 split.

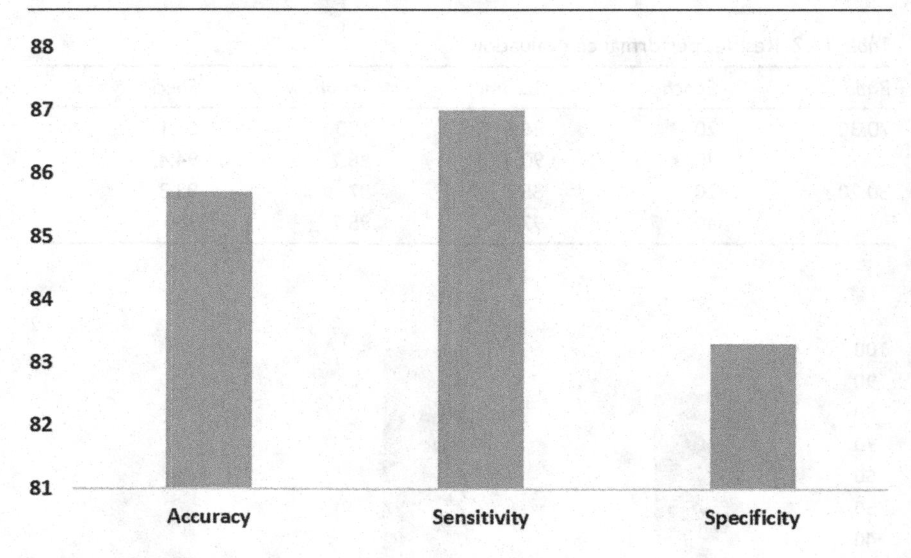

Figure 11.9 ResNet performance with 20 epochs, 80:20 split.

and Figure 11.10 show the performance of ResNet with 80:20 data split at 20 and 40 epochs, respectively.

The performance evaluation of these models attained >90% at 80:20 split and 40 epochs. AlexNet attained 97.1% accuracy, 100% sensitivity and 91.7% specificity. The AlexNet model in this study outperformed all the models used by Goyal et al. [19] for the classification of DFU. Particularly, AlexNet, in their study, achieved 89.3% accuracy, 89.5% sensitivity and 88.6% specificity; while the DFUNet they developed achieved 92.5% accuracy, 93.4% sensitivity and 91.1% specificity which are all lower than our AlexNet model achievement.

The higher achievement in this study could be attributed to the epoch and learning rate used. Our result showed that AlexNet can perform well and also outperform other models in the detection of DFU by adjusting some parameters. The 100% sensitivity in this study is in line with the result of Amin et al. [3] when Softmax and ensemble were used as classifiers in the YOLOv2-DFU models.

The ResNet-101 optimum performance in this study was attained at 80:20 split, 40 epochs of 97.1% accuracy, 95.7% sensitivity and 100% specificity. This study is the first study to employ ResNet-101 for the classification of foot images to detect DFU, and the performance is good. The 100% specificity achieved by this model is in line with the specificity obtained by Amin et al. [3] when KNN classifier was used in the YOLOv2-DFU models for DFU detection.

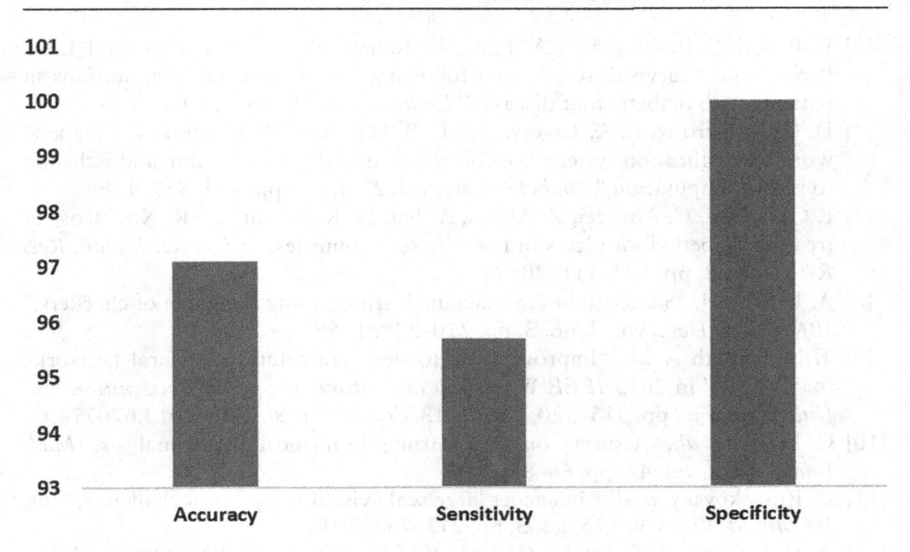

Figure 11.10 ResNet performance with 40 epochs, 80:20 split.

11.4 CONCLUSION

The performance evaluation of these models shows that both ResNet and AlexNet attained performance greater than 90% at 80:20 split and 40 epochs. These results proved that the best performance of a model can be attained at different data split ratio and also when the training parameters are fine-tuned. These models will be helpful in DFU diagnosis. In the future, we will consider the effect of data volume and try to use feature extraction techniques.

REFERENCES

[1] N. H. Cho *et al.*, "IDF diabetes atlas-8th," *Int. Diabetes Fed.*, p. 160, 2015.

[2] S. Wild, G. Roglic, A. Green, R. Sicree, and H. King, "Global prevalence of diabetes: estimates for the year 2000 and projections for 2030," *Diabetes Care*, vol. 27, no. 5, pp. 1047–1053, 2004.

[3] J. Amin, M. Sharif, M. A. Anjum, H. U. Khan, M. S. A. Malik, and S. Kadry, "An integrated design for classification and localization of diabetic foot ulcer based on CNN and YOLOv2-DFU models," *IEEE Access*, vol. 8, pp. 228586–228597, 2020.

[4] D. W. Florencia Aguirre, Alex Brown, Niam Han Cho, Gisela Dahlquist, Sheree Dodd, Trisha Dunning, Michael Hirst, Christopher Hwang, Dianna Magliano, Chris Patterson, Courtney Scott, Jonathan Shaw, Gyula Soltesz, and Juliet Usher-Smith, *IDF Diabetes Atlas: Sixth edition*, vol. 76, no. 7. International Diabetes Federation, 2013. https://research.sahmri.org.au/en/publications/idf-diabetes-atlas-sixth-edition [Online].

[5] B. P. Soo, S. Rajbhandari, A. Egun, U. Ranasinghe, I. M. Lahart, and J. M. Pappachan, "Survival at 10 years following lower extremity amputations in patients with diabetic foot disease.," *Endocr.*, vol. 69, no. 1, 2020.

[6] D. G. Armstrong, L. A. Lavery, and L. B. Harkless, "Validation of a diabetic wound classification system: the contribution of depth, infection, and ischemia to risk of amputation," *Diabetes Care*, vol. 21, no. 5, pp. 855–859, 1998.

[7] P. Cavanagh, C. Attinger, Z. Abbas, A. Bal, N. Rojas, and Z.-R. Xu, "Cost of treating diabetic foot ulcers in five different countries," *Diabetes. Metab. Res. Rev.*, vol. 28, pp. 107–111, 2012.

[8] A. L. Samuel, "Some studies in machine learning using the game of checkers," *IBM J. Res. Dev.*, vol. 3, no. 3, pp. 210–229, 1959.

[9] T. N. Sainath *et al.*, "Improvements to deep convolutional neural networks for LVCSR," in *2013 IEEE Workshop on Automatic Speech Recognition and Understanding*, pp. 315–320, Dec. 2013. doi: 10.1109/ASRU.2013.6707749.

[10] G. Litjens *et al.*, "A survey on deep learning in medical image analysis," *Med. Image Anal.*, vol. 42, pp. 60–88, 2017.

[11] O. Russakovsky *et al.*, "Imagenet large scale visual recognition challenge," *Int. J. Comput. Vis.*, vol. 115, no. 3, pp. 211–252, 2015.

[12] A. U. Ibrahim, E. Guler, M. Guvenir, K. Suer, S. Serte, and M. Ozsoz, "Automated detection of Mycobacterium tuberculosis using transfer learning," *J. Infect. Dev. Ctries.*, vol. 15, no. 05, pp. 678–686, 2021.

[13] Z. S. Ameen, M. Ozsoz, A. S. Mubarak, F. Al Turjman, and S. Serte, "C-SVR crispr: prediction of CRISPR/Cas12 guideRNA activity using deep learning models," *Alexandria Eng. J.*, vol. 60, no. 4, pp. 3501–3508, 2021, doi:10.1016/j.aej.2021.02.007.

[14] A. S. Mubarak, S. Serte, F. Al-Turjman, Z. S. Ameen, and M. Ozsoz, "Local binary pattern and deep learning feature extraction fusion for COVID-19 detection on computed tomography images," *Expert Syst.*, no. September, pp. 1–13, Sep. 2021, doi:10.1111/exsy.12842.

[15] S. Z. Amee, A. S. Mubarak, A. Süleyman, and O. Mehmet, "Development of CNN model for prediction of CRISPR/Cas12 guide RNA activity," in *Advances in intelligent systems and computing*, vol. 1095 AISC, 2020, pp. 697–703, doi:10.1007/978-3-030-35249-3_90.

[16] R. A. A. Mubarak Auwalu Saleh, Sertan Serte Fadi Al-Turjman Zubaida Sa' id Ameen and Mehmet OzsozAbdulkadir, "Deep learning-based feature extraction coupled with multi class SVM for COVID-19 detection in the IoT era Mubarak Auwalu Saleh and Sertan Serte Fadi Al-Turjman Zubaida Sa' id Ameen and Mehmet Ozsoz," *Int. J. Nanotechnol.*, vol. 1, no. 1, 2021, doi:10.1504/IJNT.2021.10040115.

[17] A. S. Mubarak *et al.*, "Metro-environmental data approach for the prediction of chemical oxygen demand in new nicosia wastewater treatment plant," *Desalin. Water Treat.*, vol. 221, no. May, pp. 31–40, 2021, doi:10.5004/dwt.2021.27049.

[18] L. Alzubaidi, M. A. Fadhel, S. R. Oleiwi, O. Al-Shamma, and J. Zhang, "DFU_QUTNet: diabetic foot ulcer classification using novel deep convolutional neural network," *Multimed. Tools Appl.*, vol. 79, no. 21, pp. 15655–15677, 2020.

[19] M. Goyal, N. D. Reeves, S. Rajbhandari, N. Ahmad, C. Wang, and M. H. Yap, "Recognition of ischaemia and infection in diabetic foot ulcers: Dataset and techniques," *Comput. Biol. Med.*, vol. 117, p. 103616, 2020.

[20] I. Cruz-Vega, D. Hernandez-Contreras, H. Peregrina-Barreto, J. de J. Rangel-Magdaleno, and J. M. Ramirez-Cortes, "Deep learning classification for diabetic foot thermograms," *Sensors*, vol. 20, no. 6, p. 1762, 2020.
[21] L. Alzubaidi, A. A. Abbood, M. A. Fadhel, O. Al-Shamma, and J. Zhang, "Comparison of hybrid convolutional neural networks models for diabetic foot ulcer classification," *J. Eng. Sci. Technol.*, vol. 16, no. 3, pp. 2001–2017, 2021.
[22] B. Cassidy *et al.*, "The DFUC 2020 dataset: Analysis towards diabetic foot ulcer detection," *touchREVIEWS Endocrinol.*, vol. 17, no. 1, p. 5, 2021.
[23] L. Torrey and J. Shavlik, "Transfer learning," in *Handbook of research on machine learning applications and trends: algorithms, methods, and techniques*, IGI global, 2010, pp. 242–264.
[24] N. Tajbakhsh *et al.*, "Convolutional neural networks for medical image analysis: Full training or fine tuning?," *IEEE Trans. Med. Imaging*, vol. 35, no. 5, pp. 1299–1312, 2016, doi:10.1109/TMI.2016.2535302.
[25] A. Krizhevsky, I. Sutskever, and G. E. Hinton, "Imagenet classification with deep convolutional neural networks," *Commun. ACM*, vol. 60, no. 6, 2017, pp. 84–90.
[26] S. Ahuja, B. K. Panigrahi, N. Dey, V. Rajinikanth, and T. K. Gandhi, "Deep transfer learning-based automated detection of COVID-19 from lung CT scan slices," *Appl. Intell.*, vol. 51, no. 1, pp. 571–585, 2021.
[27] A. M. Romano and A. A. Hernandez, "Enhanced deep learning approach for predicting invasive ductal carcinoma from histopathology images," in *2019 2nd International Conference on Artificial Intelligence and Big Data (ICAIBD)*, May 2019, pp. 142–148. doi: 10.1109/ICAIBD.2019.8837044.
[28] A. U. Ibrahim, M. Ozsoz, S. Serte, F. Al-Turjman, and P. S. Yakoi, "Pneumonia classification using deep learning from chest X-ray images during COVID-19," *Cognit. Comput.*, pp. 1–13, 2021.
[29] K. Almezhghwi, S. Serte, and F. Al-Turjman, "Convolutional neural networks for the classification of chest X-rays in the IoT era," *Multimed. Tools Appl.*, pp. 1–15, 2021.
[30] S. Serte and H. Demirel, "Wavelet-based deep learning for skin lesion classification," *IET Image Process.*, vol. 14, no. 4, pp. 720–726, 2019.
[31] Y. Celik, M. Talo, O. Yildirim, M. Karabatak, and U. R. Acharya, "Automated invasive ductal carcinoma detection based using deep transfer learning with whole-slide images," *Pattern Recognit. Lett.*, vol. 133, pp. 232–239, 2020.
[32] S. Serte, A. Serener, and F. Al-Turjman, "Deep learning in medical imaging: A brief review," *Trans. Emerg. Telecommun. Technol.*, p. e4080, 2020.
[33] X. Han, Y. Zhong, L. Cao, and L. Zhang, "Pre-trained alexnet architecture with pyramid pooling and supervision for high spatial resolution remote sensing image scene classification," *Remote Sens.*, vol. 9, no. 8, p. 848, 2017.

Chapter 12

A case study–based analysis on remote medical monitoring with AWS cloud and internet of things (IoT)

Gokul H, Atharva Deshmukh, Shraddha Jathar, and Amit Kumar Tyagi

CONTENTS

12.1 Introduction 194
12.2 Diagnosis 194
12.3 Surgery 196
12.4 Treatment 197
12.5 Covid-19 198
  12.5.1 Diagnosis 198
  12.5.2 Treatment 199
  12.5.3 Prevention 199
12.6 Other works 199
12.7 AWS system for real-time remote patient health monitoring 200
12.8 Health monitoring system requirements 201
  12.8.1 Communication between sensors and gateways 201
  12.8.2 Solution for cloud computing 203
  12.8.3 Definition of needs 203
12.9 IoT healthcare applications 204
  12.9.1 Cloud integration 204
  12.9.2 Big data in IoT healthcare 204
  12.9.3 Security in IoT 206
12.10 Challenges in healthcare IoT 206
12.11 Conclusion 207
12.12 Acknowledgement 208
12.13 Conflict of interest 208
References 208

DOI: 10.1201/9781003305347-12

12.1 INTRODUCTION

Computer vision (CV) is one area in which humans have made great advancements in the 21st century. In simple terms, it uses the camera to capture images and make meaningful interpretations from them. It has artificial intelligence and deep learning techniques of its own, and many work uses deep learning techniques with computer vision to achieve greater results. In this research paper, we explore various areas in healthcare and analyse how specific problems are solved by their application. If we look at the content covered section-wise, Section 12.2 covers topics related to new CV-based diagnosis approaches and how CV improves existing diagnosis methods. Section 12.3 covers advancements in the field of surgery where CV ensures the safety of patients. Section 12.4 analyses various works done in treating medical conditions. Section 12.5 is dedicated to advancements related to Covid-19. Section 12.6 covers a few miscellaneous works done in the field. Section 12.7 concludes the paper.

12.2 DIAGNOSIS

A cure for a disease can only be effective when it is applied to the right illness. Diagnosis forms the base of healthcare. A lot of technological advancements are happening in the field, and computer vision contributes to improving the accuracy and process of diagnosis. In this section, we explore how facial morphology can show symptoms of illness followed by stress detection and diagnostic advancements in specific diseases like dementia and cancer.

As healthcare systems become more advanced, monitoring a patient outside the hospital has become a necessity. Computer vision not only does automatic medical diagnosis but can also provide an objective secondary second assessment which can help medical personnel reach more solid conclusions and, thus, provide proper medical care. The latter one can be considered the best option now since it would be a chance for machine learning algorithms to get experience with more training data. A variety of symptoms are identified physically, and research shows that the face alone can show symptoms of up to 30 medical conditions [1]. Along with normal imaging techniques, new techniques, like thermal scanning and stereo photogrammetry, can help us see visual cues and symptoms which are otherwise not diagnosable.

Research shows that Parkinson's disease could cause abnormal eye movement [3], hepatitis could cause yellow face [4], and autism and depression could cause disturbances in facial expression [5]. Facial muscular contraction can represent pain which might be helpful for doctors to monitor a patient. These symptoms, if identified earlier, can produce significant results in cure and recovery. The advantage of computers over humans in detecting visual cues is their ability to see beyond the visible region of the spectrum. Thermal imaging can identify psychiatric, respiratory and ophthalmological conditions. The stereophotogrammetry method can detect facial paralysis and morphological abnormalities [1]. Bell's palsy can be detected using Microsoft Kinect [1].

Stress can lead to a decrease in the quality of life and other medical conditions. Detecting stress levels in daily life and taking measures to balance it can be beneficial in the long run. A study finds that computer vision can be used to predict stress levels by tracking blink rate, head motion, and eyebrow-raising and lowering [8]. Muscle hyperactivity can also be detected in videos which indicate an increase in stress levels.

Computer vision and imaging techniques can be applied to medical selfies that can then be processed to identify various diseases like jaundice and prostate cancer. Jaundice is a health condition caused by a high level of bilirubin in the blood. Infants are most affected by this. The yellow discolouration is one identifiable symptom. Phototherapy is proven to be an effective cure for this. In a study, infants were analysed by an HD camera [6] which initially does colour transformations in the region of interest. It then analyses and decides whether phototherapy is required or not. An SMS is sent, and the machine is turned on or off accordingly.

Jaundice is a symptom of prostate cancer, and detection of jaundice would help to detect and to cure prostate cancer at an earlier stage. BiliScreen is an app that detects the bilirubin levels of a person by analysing their eye [7]. It uses two devices to calibrate and capture an image which is then processed to measure the bilirubin level. This app comes to the rescue in diseases like prostate cancer, which has jaundice as its preliminary stage. This app tracks even minor changes not noticeable by the human eye, thus enabling physicians to make better decisions and early diagnoses.

Misinterpretation accounts for 10–30% in diagnosis errors [38]. Radiologists can make mistakes by identifying malignant cancer as benignant or can approve false-negative results. Prognosis is as important as the diagnosis. A computer-aided diagnosis (CAD) [9] system can improve the accuracy of diagnosis. It uses deep learning combined with computer vision to perform segmentation and feature extraction. Many conditions, like dementia, pneumonia, cancer and Covid-19, can be more accurately diagnosed with computer-aided diagnosis. Condition-specific methods are discussed subsequently.

Alzheimer's disease can be detected at an early stage through MRI scans. A study [39] proposes a transfer learning–based method for analysing structural MRI images. Research shows that machines do a better job predicting Alzheimer's disease than clinicians [40].

Breast cancer detection is mostly done by breast mammogram screening methods. Its accuracy is low in detecting cancer in dense breasts, especially in younger women [41]. Computer-aided design (CAD) uses image segmentation to identify suspicious areas which are then confirmed by physicians. It classifies abnormalities as benign or malignant. Computer vision techniques used include texture-based segmentation, low-level thresholding and mathematical morphology [41]. Techniques using convolutional neural networks (CNN) show significantly higher results than those using statistical methods.

Ultrasound is commonly used to diagnose prostate cancer, and early detection may improve the survival rate. Noise in ultrasound makes it difficult to perform classical automatic segmentation. 3D segmentation, along with CAD, can detect prostate boundaries and guide the biopsy system [41].

In the case of lung cancer, lung nodules are detected through a computed tomography (CT) scan. Rib cage and air pockets interfere with the detection of lung nodules. A proposed system [41] uses stereoscopic visualization and background blurring for nodule detection cards, and supervised learning methods further decrease the identification of false positives.

In skin cancer, colour quantization and region-growing algorithms are proven to detect malignant melanoma [42], which accounts for the majority of skin cancer mortality.

AI-driven digital histopathology is a new field that has emerged due to advancements in the field of computer vision. Humans are limited to a certain level of visual perception which can sometimes lead to inconsistencies in diagnosis. Computers can improve the efficiency of such routine tasks and can also analyse the morphological structure invisible to the human eye. One example is the localization quantification of morphological features, like cells, nuclei and mitosis [43].

Computer vision and deep learning have been able to diagnose dozens of skin conditions by using techniques like lesion-specific differential diagnosis. In ophthalmology, methods like optical coherence tomography (OCT), along with computer vision algorithms, were able to diagnose conditions like glaucoma and childhood blindness [43].

A lot of work that is mentioned earlier is built on top of the existing systems. Such an approach helps the widespread adoption of these technologies without much cost overhead. It can save professionals time, and doctors can focus their time on people who require urgent attention.

12.3 SURGERY

Computer vision can help ensure additional patient safety in operating rooms during surgery. Humans are prone to error and might lose focus when doing a job repeatedly or for a long time. This section explores how surgical video analysis can play an important role followed by its impact on minimally invasive surgery and open-heart surgery.

New technology makes operating rooms complex, and surgical safety is currently implemented by trained human observers who collect data manually. Computer vision tools can be trained and used to collect data that avoid human errors and ensure privacy. It can also analyse how healthcare professionals perform their job or check for any protocol breach [13]. Research shows the real-time tracking of the operating environment during a laparoscopic cholecystectomy [33]. It measures situational awareness and other factors in real time. A multi-source acquisition system identifies intraoperative distractions [35]. The Hawthorne effect clearly states that the behaviour changes can be situational and do not reflect typical behaviour [34]; even considering this, the system takes care of the safety concerns. In some cases, it can detect and notify any bleeding that is unattended. Another research study implements an analysis of situational awareness [36]. CONDOR (Connected Optimised Network and Data in Operating Rooms) [13] and Black Box (Surgical Safety Technologies Inc., Toronto, Ontario) [33] are two real-world implementations of the previously mentioned technology.

Minimally invasive surgery (MIS) is often preferred over normal surgery as it reduces post-recovery time and causes less trauma. Computer vision helps MIS achieve its true potential. Here, we discuss laparoscopic MIS. Recent innovations in computer vision can help construct a 3D viewpoint of the scene. A lot of visual information is lost when a scene from a 3D world is represented in 2D. Computer vision algorithms, when trained on preoperative data, were able to even detect malignant tumours which could be hidden from a surgeon's viewpoint [22].

Computer vision is also used in live open-heart surgery to track surgical motion with minimum inconvenience to surgeons [37]. The camera is mounted above the operating table.

12.4 TREATMENT

As the life expectancy grows higher, more people would need access to healthcare facilities. Public hospitals receive a huge number of patients and may not be able to give attention to everyone. The traditional methods involve face-to-face assessment done periodically which requires a professional's presence. With the help of computer vision, a team in Malaysia has developed a low-cost patient monitoring system [14]. It has an emotional module that continuously analyses the facial expressions of the patient and predicts emotion levels. It also has a vital module that includes an array of sensors. When the values go beyond the limits, a notification is sent to the medical personnel. This helps unattended patients get medical care in case of an emergency.

Mental health is as important as physical health and has to be treated when needed. Research shows that more than 70% of cases are not treated or brought to attention [11]. Cognitive impairment is one such situation. Cognitive impairment can be defined as one's inability to perform cognitive functions like thinking, memorizing and reasoning [12]. Early diagnosis and constant follow-ups are needed in many cases for effective treatment. Digital biomarkers are one way to measure cognitive health. They are objective, quantifiable data collected by devices that can analyse psychological and behavioural patterns. Meanwhile, the individuals must also feel comfortable using the devices. Digital games are one such activity where the user has to perform a cognitive function at the same time to feel comfortable. Researchers have been able to identify and incorporate 10 digital biomarkers into the Microsoft Solitaire Collection which can then be processed by an open-source computer vision library [10]. This technology has great potential as millions of people suffer from these conditions, and early detection through easy and cost-effective methods could make an impact.

Medication helps cure illness and plays an important role in healthcare. One of the challenges faced in this area is the use of correct medicine in prescribed doses, and any change might result in death. A system is proposed to verify medicine dispensing using an attentive computer vision approach [15]. One can enter the patient ID into the machine which can automatically retrieve the prescription details from the database, and the pills in the dispensing cup can be cross-verified. If there is any mismatch, the system alerts the user of the change. It uses a pill descriptor vector (PDV) which contains data, enough to identify pills uniquely. It contains planar geometry components, colour and infrared components, and surface intensity descriptors. The new dataset using rotation-invariant image descriptors further helps with the identification.

12.5 COVID-19

The SARS-CoV-2 or Covid-19 pandemic has hit humanity very hard. Researchers around the globe contributed a lot of resources to fight Covid-19. This time has been dubbed as 'Science of the times' [27] for the enormous and quick response of the science community. In this section, we discuss how computer vision helps tackle problems related to the diagnosis, treatment and prevention of Covid-19.

12.5.1 Diagnosis

Diagnosis of Covid-19 traditionally uses techniques like reverse transcriptase quantitative polymerase chain reaction (RT-qPCR). While tests like RT-qPCR are proven accurate, it is expensive, time-consuming and manual. With an ever-increasing number of patients, it is not feasible for everyone to undergo these tests. Computed tomography (CT) scans are also used to confirm Covid-19, but it is limited to the ability of a medical professional to identify abnormalities in the result to confirm it. Computer vision can help doctors make a better prognosis. Three classes of work have mainly been found, and includes using UNet++ semantic segmentation model [31], the VNET-based segmentation model [32] and the binary classification model [30]. UNet++ semantic segmentation model identifies infected areas of a person's scan from a healthy ones. This model has been trained and tested with a 46,096 CT image dataset and has shown significant positive results. It was deployed at the Renmin Hospital of Wuhan University (Wuhan, Hubei province, China) [29]. This was the only model which was applied in real life.

X-ray is a low-cost, widely available solution. COVID-Net, proposed by Darwin AI, enables the detection of Covid-19 from chest X-rays [21]. It can classify a scan as normal infection, non-Covid-19 infection and Covid-19 infection. It has achieved 92.4% accuracy in tests. Other techniques were also developed using various neural networks and deep learning techniques [20].

12.5.2 Treatment

Although no treatment has been found for Covid-19 to date. Computer vision and deep learning algorithms can help classify patients according to the severity of their illnesses. A respiratory simulation model (RSM) can classify several respiratory patterns with great accuracy and precision. It is then used to calculate a disease progression score called the corona score. Patients can be classified based on this corona score.

12.5.3 Prevention

It is time-tested that prevention is better than cure [28]. A lot of effort has been put into developing technologies that prevent the spread of Covid-19. Computer vision algorithms are used in China to verify if people are wearing mask [20]. Infrared thermography is also used in public places to screen people with fever and can replace one-on-one manual scanning [23]. Drones were also equipped to scan mass crowds. Germ identification from microscopic datasets was proposed by Edouard A. Hay [24]. Other technologies like controlling elevators and doors use computer vision to prevent transmission of Covid-19 through surfaces by touching [25]. In Jakarta, YOLO and MobileNet SSD were used to ensure social distancing and object counting through CCTV cameras installed around the city [26].

12.6 OTHER WORKS

Healthcare is about the prevention, treatment and diagnosis of diseases. Supporting people with disabilities is also a part of healthcare. Wearable devices are now capable of tracking various parameters of our health, like pulse rate. Research [19] takes wearable devices and does palm recognition, thus enabling us to analyse gestures, fingertip movements and so on. This can form the foundation for user interfaces for people with disabilities.

Another research [18] tracks upper body joints, performs spatiotemporal segmentation, performs hand gesture recognition and performs human pose estimation to understand sign language. It enables people who use sign language (American Sign Language) to communicate through video, which can be automatically interpreted. Although the research is now limited to static images, real-time programs are expected to be developed soon.

This research [17] takes a completely different approach by trying to mimic human vision using computer vision and deep learning. Initially, the brain activity of subjects is recorded using 28 channel EEG (electroencephalography) while they look at various image classes. This helps analyse how visual classification is done by the human brain. Then a computer vision system is deployed to classify visual descriptors extracted from the understanding of human neural processes. Both systems combined try to transfer the visual capability of humans into machines.

Human activity recognition is a recently developed field in computer vision, and a research paper [16] explores various methods for tracking and analysing human behaviour. It also discusses various challenges and limitations of these methods.

E-Sanjeevani is a real-world application that focuses on integrating medical selfies into electronic health records (EHR), which can be processed by computer vision algorithms to identify potentially masked diseases [2]. It uses local binary pattern LBP and multidimensional big data analysis with machine learning and computer vision algorithms to achieve it.

12.7 AWS SYSTEM FOR REAL-TIME REMOTE PATIENT HEALTH MONITORING

AWS IoT: This is a framework that allows Raspberry Pi devices to communicate safely and bi-directionally with the Amazon Web Services cloud. Based on the content of the data, the IoT service determines the next step to take: If observed healthcare parameters remain regular, then information is kept in DynamoDB and S3; otherwise, the wireless sensor service sends signals to notify that systems linked elements of an emergency.

Data exploration: For processing massive amounts of information in a distributed environment in a timely and costly manner, we use a Hadoop-based service, Elastic MapReduce. This even allows for the treatment of large amounts of data across several clusters. In the case of a medical emergency involving a patient being followed, such a system will allow medical experts to also be informed real time. It also enables artificial intelligence algorithms to analyse vast medical databases in order to extract usable information, allowing clinicians to base clinical diagnoses in part on statistical inferences offered by the computer. It also allows medical and clinical offices to avoid the difficulties that come with maintaining and managing huge IT systems.

Destination: The AWS IoT gateway sends data to the Simple Storage Service and DynamoDB databases. The purpose of the Simple Storage Service is to rapidly store and prepare all incoming data for analysis. DynamoDB, like any other database, is used to analyse data via a web application. The application is linked via Elastic Load Balancing, which handles user access traffic and distributes it among software installed using the Auto Scaling concept, as shown in Figure 12.1. These systems are part of a virtual environment called a virtual private cloud, which uses a unique IP address for high-speed data transmission.

Data producer: That's the data source and, in this example, is symbolized by a Raspberry Pi chip. This captures sensor data and transmits it to the internet using the NOD-RED programming tool. This tool is used to connect online services, APIs and hardware devices; in our system, it connects the manufacturer to the gateway so that MQ messages containing the collected data may be delivered.

One issue here is the lack of connection to the centralized server, which results in data loss. As a result, data can be saved in buffer memory on the Raspberry Pi until the connection is re-established. Finally, in an emergency, the Raspberry Pi can be outfitted with a simple decision program that will set off alarms for the patient.

Event: In the event of an emergency, it indicates the alert that the system sends to the physician. This alert makes use of Amazon SNS to send messages to iPhone or Android apps that are connected to Mobile-Hub, a service that functions as an event handler between the SNS service and the AWS IoT gateway. From a single integrated console, Mobile-Hub can monitor application usability as well as statistics and analytics on requests.

12.8 HEALTH MONITORING SYSTEM REQUIREMENTS

12.8.1 Communication between sensors and gateways

Continuous improvements within nanoelectronics in the last few generations have ensured a significant amount of component integration, culminating in the birth of microscopic electronic devices made up of micro-sensors that can gather and transmit data independently. As a result, the new Internet of Things (IoT) paradigm arose, associating items, connections and the online world, and expressing information and data exchange between these devices and the internet. Ever since, this concept of human carrying smart sensors to remotely transmit information or simply for interconnections within the

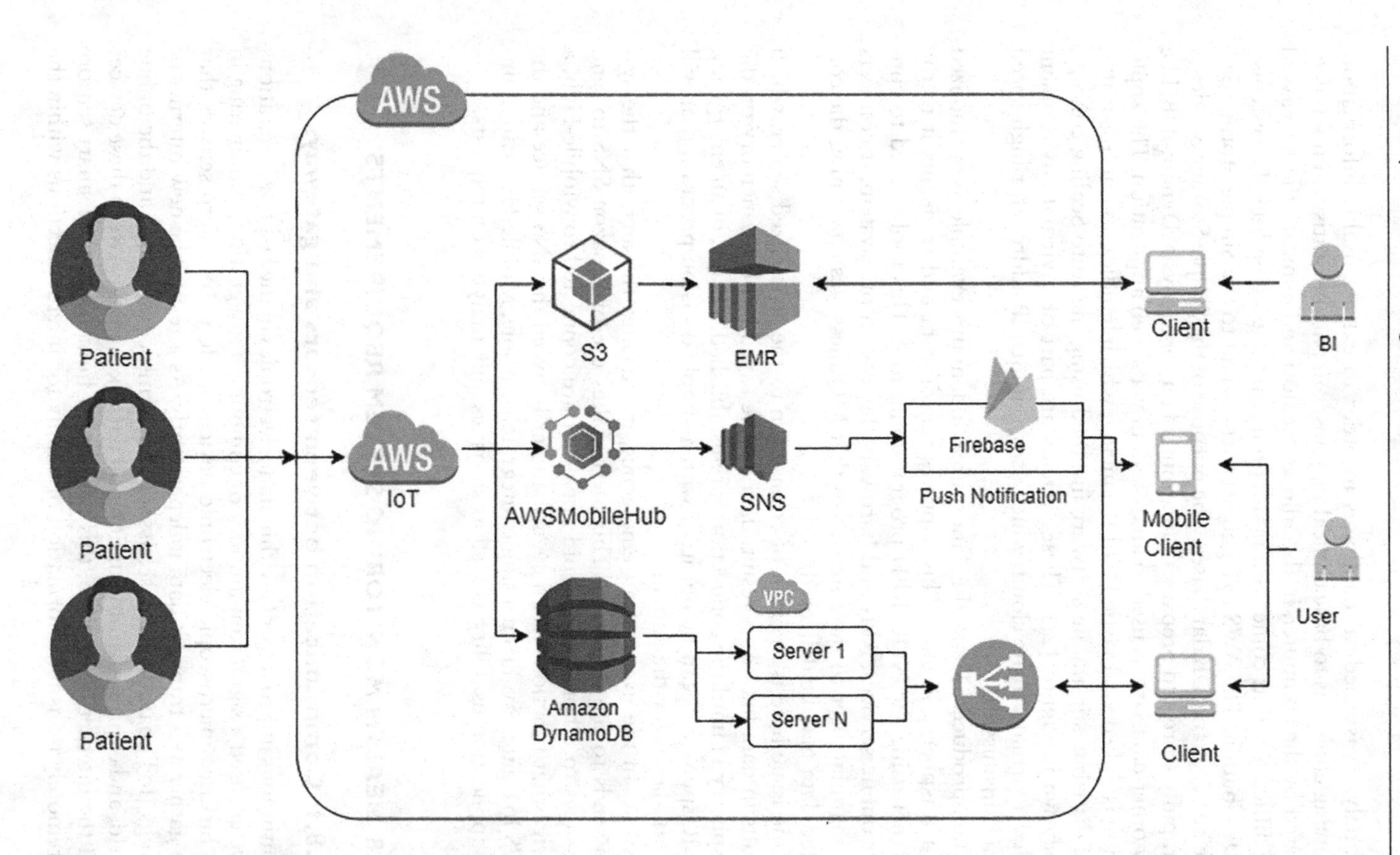

Figure 12.1 Architecture of the health monitoring system.

human have sprouted up, spawning plenty of potential applications in various areas as diverse as human-machine interaction, entertainment, health and sports, to name a few.

The three primary components of IoT communication are sensors, a server and a gateway device. The sensors for medical applications are either smartwatches or surgically implanted on the patient, and they communicate with the gateway through a short-range wireless protocol, such as Zigbee or Bluetooth. The gateway connects to the cloud server via an access network, such as 3G/4G LTE protocols, Ethernet or Wi-Fi.

12.8.2 Solution for cloud computing

The amount of data generated via IoT activities for a given application may quickly exceed the storage hardware capacity of standalone servers. That's why utilizing servers from off-site systems is becoming more common. Cloud computing is a new computer paradigm that involves offering on-demand computing resources to businesses or consumers over the internet. Infrastructure services, such as data centres for storage and processing, could be among these resources. Platform services, including the software platform for applications, can also be included. They could also be software-as-a-service apps. This method of computing helps you to save money on hardware purchases and maintenance. These servers can also be rented on demand or at a set price, depending on technical characteristics (power, bandwidth and so on). Cloud computing is particularly notable for its adaptability: users can operate their own server or just use remote programs in SaaS mode, depending on their skill level.

12.8.3 Definition of needs

The infrastructure domain is where any Internet of Things application's requirements appear first; one looks for a cloud platform which ensures transparency, dynamic scalability and total availability into the solution's system requirements.

In terms of connection, real-time remote healthcare observation of individuals is distinguished mostly by the transfer of tiny quantities of various data sources. A device connection technique is more efficient in this circumstance than the standard HTTP protocol. The new MQTT protocol has proven to be fairly useful in this field, owing to its lightness and capacity to manage a large number of deployed IoT devices, as well as its efficient communication and low-power operation.

Another important feature is information processing; in reality, data received should first undergo a specific examination to determine whether any serious conditions necessitate prompt medical intervention; alternatively, the data is sent straight into a database for ultimate analysis of data.

All these characteristics are programmable and fall under the purview of the PaaS domain.

Additionally, SaaS services, which include cloud provider APIs, can be used not only to update and manage services but also to combine the used apps in a recognizable manner for the final user.

12.9 IOT HEALTHCARE APPLICATIONS

12.9.1 Cloud integration

Remote health monitoring via smartphone application IoT data is stored in the cloud platform, which provides flexibility, scalability and greater resources for data processing. Because IoT data is collected from various sensors, it is effectively stored at a cloud-side server known as a cloud storage repository. Few researchers' medical operations are incorporated into the cloud and improved healthcare by employing cloud technology. The physiological features of students are recorded and kept in a separate manner on the cloud. When the client system completes collecting data via IoT medical equipment, it transfers it to the cloud system for analysis, according to the doctors, hospitals and caretakers' emergency messages. The research team developed a hierarchical computing architecture for a patient tracking system, which features automated information management and processing at the layer's edge.

To give a solution for network latency in healthcare processing remotely, a concept called UbeHealth is being created to investigate network delay and quality of service elements in order to improve healthcare performance in smart cities. For disease diagnosis and severity reduction, a fuzzy rule-based neural classifier has been developed. This method investigates cloud data processing by employing a secure storage mechanism, which includes terms like partitioning, merging, data retrieval and aggregation.

12.9.2 Big data in IoT healthcare

In recent years, big data storage technologies have already been important for storing enormous amounts of healthcare data. Cloud storage gets massive and is managed by a technology known as big data. According to recent research, the combination of big data as well as the cloud is having an impact on distant healthcare. Amazon Elastic MapReduce provides an innovative approach to handling large amounts of data and distributing it to a cluster. For putting data into the Hbase cluster, the Amazon EMR provides a distinct approach. Loading sensor data from Amazon S3 to HBase uses the tool as a guinea pig. Because Apache Pig is used for data processing in distributed databases, healthcare applications can significantly

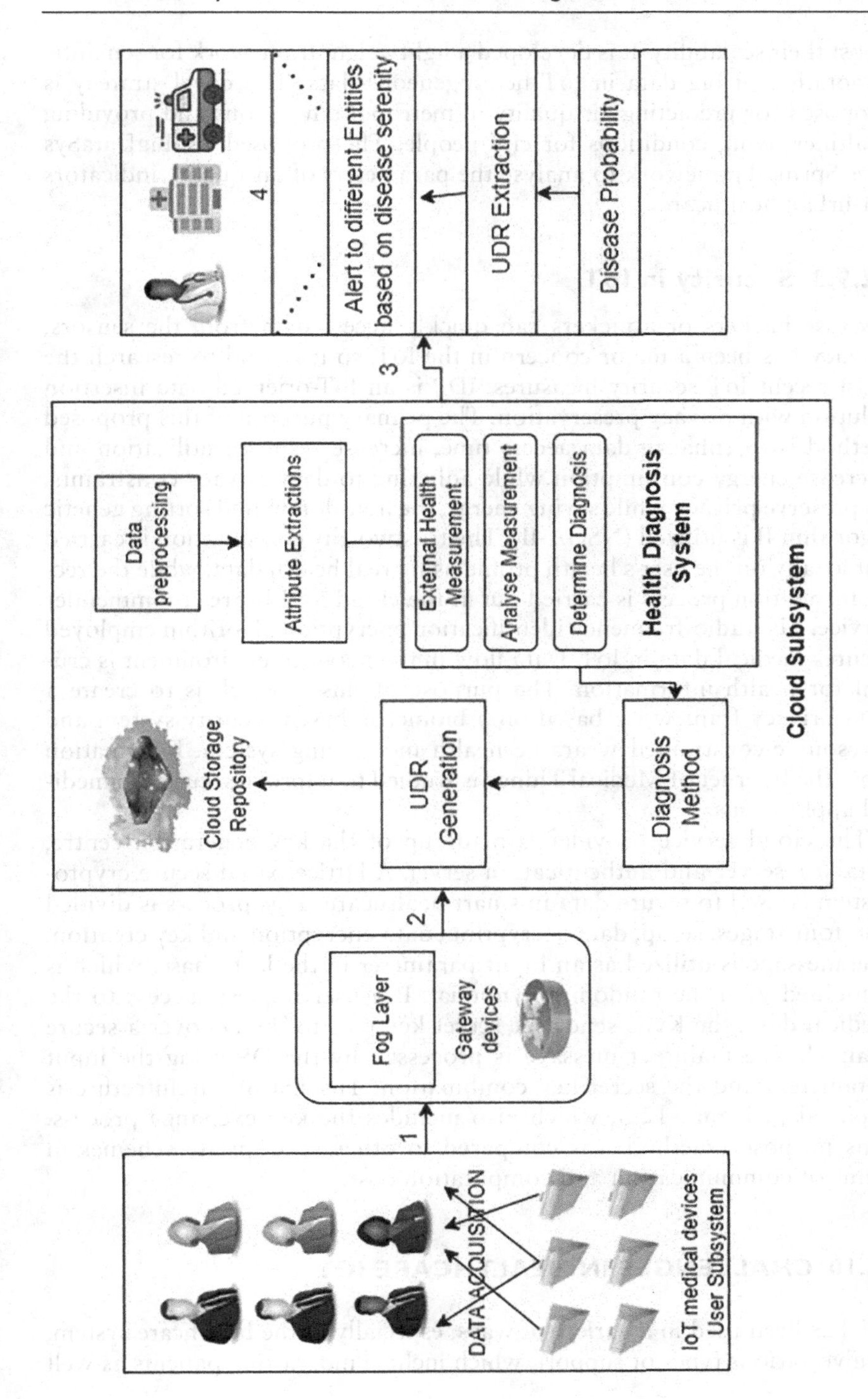

Figure 12.2 Cloud integration with healthcare IoT system.

boost their scalability. It is developed a lightweight framework for semantic annotation of big data in IoT heterogeneous data. The novel strategy is proposed for predicting air quality in metropolitan regions and providing healthier living conditions for city people. The proposed UHBigDataSys uses Spring Framework to analyse the parameters of air quality indicators for urban healthcare.

12.9.3 Security in IoT

Because hackers or attackers can quickly access data from the sensors, privacy has been a major concern in the IoT, so it is vital to research the most recent IoT security measures. IDP is an IoT-oriented data insertion solution with privacy preservation. The primary purpose of this proposed method is to enhance data access time, increase resource utilization and decrease energy consumption while adhering to data privacy constraints. To preserve privacy while saving energy, the non-dominated sorting genetic algorithm II is utilized (NSGA-II). The trustworthy computation is carried out locally on the user's health profile using real health data, while the recommendation process is carried out at the cloud healthcare recommender service. The radio-frequency identification encryption algorithm employed secures medical data in IoT. Data flow in the network environment is crucial for health information. The purpose of this research is to create a data privacy framework based on a biometric-based security system and a resource-constrained wearable health monitoring system. Information from the Internet of Medical Things is studied to improve security in medical applications.

The cloud service provider is made up of the key generation centre, database server and authentication server. A lattice-based secure cryptosystem is used to secure data in smart healthcare. This process is divided into four stages: setup, data decryption, data encryption and key creation. The message is utilized as an input parameter in the last phase, which is combined with the random polynomial. If a user requests access to the medical data, the KGC sends the secret key pair to the DS over a secure channel. The plaintext message is processed by the DS using the input parameters and the secret key combination. The LSCSH architecture is depicted in Figure 12.3, which also includes the key exchange process. This proposed method was compared to other appropriate schemes in terms of communication and computation cost.

12.10 CHALLENGES IN HEALTHCARE IOT

IoT has been used in a variety of ways, especially in the healthcare system, to give various types of support, which include monitoring patients as well as a smart home system for people with diabetes. The biggest challenges that occur in the healthcare system are listed subsequently:

- IoT enables tremendous flexibility; for example, when a patient needs continuous nursing, he or she might stay home instead of in a hospital and also be observed on a regular basis through IoT technology. Sometimes, fitness trackers, which include sensors, cause discomfort to the patient.
- Information is transmitted first from sensors to the remote controller and finally to the monitoring centre, with noise impacting quality of the data. An improved architecture enables data transport while maintaining its integrity. This data signal can also be enhanced using the noise removal technique.
- Guided signal analysis is used in the majority of current ECG monitoring systems. This increases the cost and may lead to inaccurate detection. Signal analysis can be done using machine learning, which increases productivity and lowers costs.
- As the number of sensors and devices increases, so does the amount of energy required to process them, resulting in increased power loss and consumption. To save energy, an optimization technique can be used. In the IoT, monitoring a high number of users demands extra storage and mainframe resources, which can be avoided by storing data in the cloud. Integration of IoT with the cloud, however, adds to the complexity.
- Another major issue in the IoT is privacy, which is made worse by the truth that gadgets are increasingly vulnerable to attack. Because these devices have limited resources, applying encryption techniques to them is tricky.

12.11 CONCLUSION

Computer vision plays a major role in the healthcare system. Starting from diagnosis to cure it influences almost all aspects of the system. It is clear that computer vision helps in the early detection of diseases and can help save the patient. Computer vision is used to detect symptoms from the face. Computer-aided diagnosis improved many existing diagnosing methods for better accurate prediction. Cancer is detected in its early stages. Surgical video is a rich data source and can be analysed to improve the safety of the procedure. New surgical methods like minimally invasive surgery take great advantage of computer vision and deep learning methods. Patients can now be monitored without human intervention. Mental health can now be continuously tracked. Medication can now be verified before it is taken. In the case of Covid-19, it helps us achieve more significant results in a shorter timeline and ensured safety by mass crowd temperature scanning. Beyond healthcare, facilities also help us track our daily health. It offers great relief to people with disabilities as now, they can technically overcome it. Even the human visual system has been tried to be copied into the machine. Computer vision has come a long way and now contributes significantly to our life. We expect even more contributions in the field of healthcare in the future.

12.12 ACKNOWLEDGEMENT

I would like to express sincere gratitude to my institution, Vellore Institute of Technology in Chennai, and my mentor Dr Amit Kumar Tyagi for introducing me to the idea of writing a research paper.

12.13 CONFLICT OF INTEREST

The author states that there are no conflicts of interest.

REFERENCES

[1] J. Thevenot, M. B. López, and A. Hadid, "A Survey on Computer Vision for Assistive Medical Diagnosis From Faces," *IEEE Journal of Biomedical and Health Informatics*, vol. 22, no. 5, pp. 1497–1511, Sept. 2018. http://doi.org/10.1109/JBHI.2017.2754861.

[2] Vuppalapati, J.S., Kedari, S., Kedari, S., Vuppalapati, R., Vuppalapati, C. and Ilapakurti, A., 2018, December. The role of selfies in creating the next generation computer vision infused outpatient data driven electronic health records (EHR). In *2018 IEEE International Conference on Big Data (Big Data)* (pp. 2458–2466). IEEE.

[3] P. H. Tseng, I. G. Cameron, G. Pari, J. N. Reynolds, D. P Munoz, and L. Itti, "High-Throughput Classification of Clinical Populations from Natural Viewing Eye Movements," *Journal of Neurology*, vol. 260, no. 1, pp. 275–284, 2013.

[4] M. Liu and Z. H. Guo, "Hepatitis Diagnosis Using Facial Color Image," in *Medical Biometrics*. ICMB 2008 (vol. 4901 of Lecture Notes Computer-Science). New York: Springer, 2007, pp. 160–167.

[5] Liu, T., Yuan, Z., Sun, J., Wang, J., Zheng, N., Tang, X. and Shum, H.Y., 2010. Learning to detect a salient object. *IEEE Transactions on Pattern analysis and machine intelligence*, 33(2), pp. 353–367.

[6] Montanari, D., Bremer, J., Gendotti, A., Geynisman, M., Hentschel, S., Loew, T., Mladenov, D., Montanari, C., Murphy, S., Nessi, M. and Norris, B., 2015, November. Development of membrane cryostats for large liquid argon neutrino detectors. In *IOP Conference Series: Materials Science and Engineering* (Vol. 101, No. 1, p. 012049). IOP Publishing.

[7] Mariakakis, A., Banks, M.A., Phillipi, L., Yu, L., Taylor, J. and Patel, S.N., 2017. Biliscreen: smartphone-based scleral jaundice monitoring for liver and pancreatic disorders. *Proceedings of the ACM on Interactive, Mobile, Wearable and Ubiquitous Technologies*, 1(2), pp. 1–26.

[8] D. Manousos et al., "Contactless Detection of Facial Signs Related to Stress: A Preliminary Study," *2014 4th International Conference on Wireless Mobile Communication and Healthcare – Transforming Healthcare Through Innovations in Mobile and Wireless Technologies (MOBIHEALTH)*, 2014, pp. 335–338. https://doi.org/10.1109/MOBIHEALTH.2014.7015979.

[9] C. Angelica Charleen, H. Purnama and F. Purnomo, "Impact of Computer Vision with Deep Learning Approach in Medical Imaging Diagnosis," *2021 1st International Conference on Computer Science and Artificial Intelligence (ICCSAI)*, 2021, pp. 37–41. http://doi.org/10.1109/ICCSAI53272.2021.9609708.

[10] Gielis, K., Kennes, J., De Dobbeleer, C., Puttemans, S. and Abeele, V.V., 2019, June. Collecting digital biomarkers on cognitive health through computer vision and gameplay: An image processing toolkit for card games. In *2019 IEEE International Conference on Healthcare Informatics (ICHI)* (pp. 1–12). IEEE.

[11] C. Henderson, S. Evans-Lacko and G. Thornicroft, "Mental Illness Stigma, Help Seeking, and Public Health Programs," *American Journal of Public Health*, vol. 103, no. 5, pp. 777–780, 2013. http://doi.org/10.2105/AJPH.2012.301056.

[12] E. Roy, "Cognitive Impairment," in Gellman M.D., Turner J.R. (eds) *Encyclopedia of Behavioral Medicine*. New York: Springer, 2013. https://doi.org/10.1007/978-1-4419-1005-9_1118.

[13] L. R. Kennedy-Metz et al., "Computer Vision in the Operating Room: Opportunities and Caveats," *IEEE Transactions on Medical Robotics and Bionics*, vol. 3, no. 1, pp. 2–10, Feb. 2021. http://doi.org/10.1109/TMRB.2020.3040002.

[14] Tawafak, R.M., Alfarsi, G., Malik, S.I., Eldow, A., Jabbar, J. and Alsideiri, A., 2020, September. Cancer Treatment Disease Application Software Technology in Medicine. In *2020 IEEE Student Conference on Research and Development (SCOReD)* (pp. 380–384). IEEE.

[15] Pal, C., Pankaj, S., Akram, W., Acharyya, A. and Biswas, D., 2018, May. Modified Huffman based compression methodology for deep neural network implementation on resource constrained mobile platforms. In *2018 IEEE International Symposium on Circuits and Systems (ISCAS)* (pp. 1–5). IEEE.

[16] Zhiqiang Wei, Jie Nie, Lei Huang, Shuang Wang and Zhen Li, "A Review on Human Activity Recognition Using Vision-Based Method," *Journal of Healthcare Engineering*, pp. 1–31, 2017. http://doi.org/10.1155/2017/3090343.

[17] Spampinato, C., Palazzo, S., Kavasidis, I., Giordano, D., Souly, N. and Shah, M., 2017. Deep learning human mind for automated visual classification. In *Proceedings of the IEEE Conference on Computer Vision and Pattern Recognition* (pp. 6809–6817).

[18] Gattupalli, S., 2015, October. Human motion analysis and vision-based articulated pose estimation. In *2015 International Conference on Healthcare Informatics* (pp. 470–470). IEEE.

[19] F. R. Hsu, Y. -H. Kuo, S. -Y. Wei, Y. -H. Hsieh and D. C. Nguyen, "A Study of User Interface with Wearable Devices Based on Computer Vision," *IEEE*

Consumer Electronics Magazine, vol. 9, no. 1, pp. 43–48, 1 Jan. 2020. http://doi.org/10.1109/MCE.2019.2941463.

[20] A. Ulhaq, J. Born, A. Khan, D. P. S. Gomes, S. Chakraborty and M. Paul, "COVID-19 Control by Computer Vision Approaches: A Survey," *IEEE Access*, vol. 8, pp. 179437–179456, 2020. http://doi.org/10.1109/ACCESS.2020.3027685.

[21] L. Wang, Z. Q. Lin and A. Wong, "COVID-Net: A Tailored Deep Convolutional Neural Network Design for Detection of COVID-19 Cases from Chest X-ray Images," *Scientific Reports*, vol. 10, p. 19549, 2020. https://doi.org/10.1038/s41598-020-76550-z.

[22] A. Bartoli, T. Collins, N. Bourdel and M. Canis, "Computer Assisted Minimally Invasive Surgery: Is Medical Computer Vision the Answer to Improving Laparosurgery?" *Med Hypotheses*, vol. 79, no. 6, pp. 858–863, 2012. https://doi.org/10.1016/j.mehy.2012.09.007.

[23] B. B. Lahiri, S. Bagavathiappan, T. Jayakumar and John Philip, "Medical Applications of Infrared Thermography: A Review," *Infrared Physics & Technology*, vol. 55, no. 4, pp. 221–235, 2012.

[24] Edouard A. Hay and Raghuveer Parthasarathy, "Performance of Convolutional Neural Networks for Identification of Bacteria in 3d Microscopy Datasets," *PLoS Computational Biology*, vol. 14, no. 12, p. e1006628, 2018.

[25] J. Katti, A. Kulkarni, A. Pachange, A. Jadhav and P. Nikam, "Contactless Elevator Based on Hand Gestures During Covid 19 Like Pandemics," *2021 7th International Conference on Advanced Computing and Communication Systems (ICACCS)*, 2021, pp. 672–676. https://doi.org/10.1109/ICACCS51430.2021.9441827.

[26] M. Lanang Afkaar Ar et al., "A Computer Vision-based Object Detection and Counting for COVID-19 Protocol Compliance: A Case Study of Jakarta," *2020 International Conference on ICT for Smart Society (ICISS)*, 2020, pp. 1–5. https://doi.org/10.1109/ICISS50791.2020.9307594.

[27] "Coronavirus Research is Being Published at a Furious Pace," https://www.economist.com/graphic-detail/2020/03/20/coronavirus-research-is-being-published-at-a-furiouspace, accessed February 24, 2022.

[28] L. Borysiewicz, "Prevention Is Better Than Cure," *Clinical Medicine (Lond)*, vol. 9, no. 6, pp. 572–583, 2009. https://doi.org/10.7861/clinmedicine.9-6-572.

[29] Jun Chen, Lianlian Wu, Jun Zhang, Liang Zhang, Dexin Gong, Yilin Zhao, Shan Hu, Yonggui Wang, Xiao Hu, Biqing Zheng et al. "Deep Learning-based Model for Detecting 2019 Novel Coronavirus Pneumonia on High-resolution Computed Tomography: A Prospective Study," *medRxiv*, 2020.

[30] Li, Lin, Lixin Qin, Zeguo Xu, Youbing Yin, Xin Wang, Bin Kong, Junjie Bai et al. "Artificial intelligence distinguishes COVID-19 from community acquired pneumonia on chest CT." *Radiology* (2020).

[31] Zongwei Zhou, Md Mahfuzur Rahman Siddiquee, Nima Tajbakhsh and Jianming Liang, "Unet++: A Nested U-net Architecture for Medical Image Segmentation," in *Deep Learning in Medical Image Analysis and Multimodal Learning for Clinical Decision Support*, pp. 3–11. Springer, Cham, 2018.

[32] E. Gibson et al., "Automatic Multi-Organ Segmentation on Abdominal CT with Dense V-Networks," *IEEE Transactions on Medical Imaging*, vol. 37, no. 8, pp. 1822–1834, Aug. 2018. https://doi.org/10.1109/TMI.2018.2806309.

[33] S. Guerlain et al., "Assessing Team Performance in the Operating Room: Development and Use of a 'Black-box' Recorder and Other Tools for the Intraoperative Environment," *Journal of the American College of Surgeons*, vol. 200, no. 1, pp. 29–37, 2005.
[34] J. McCambridge, J. Witton, and D. R. Elbourne, "Systematic Review of the Hawthorne Effect: New Concepts Are Needed to Study Research Participation Effects," *Journal of Clinical Epidemiology*, vol. 67, no. 3, pp. 267–277, 2014. https://doi.org/10.1016/j.jclinepi.2013.08.015.
[35] M. G. Goldenberg, J. Jung, and T. P. Grantcharov, "Using Data to Enhance Performance and Improve Quality and Safety in Surgery," vol. 152, no. 10, pp. 972–973, 2017.
[36] Bardram, Jakob E., and Niels Nørskov. "A context-aware patient safety system for the operating room." *Proceedings of the 10th International Conference on Ubiquitous Computing*. 2008.
[37] M. Stenmark, E. Omerbašić, M. Magnusson, V. Andersson, M. Abrahamsson, and P. K. Tran, "Vision-Based Tracking of Surgical Motion During Live Open-Heart Surgery," *Journal of Surgical Research*, vol. 271, pp. 106–116, 2022. https://doi.org/10.1016/j.jss.2021.10.025.
[38] J. Gao, Q. Jiang, B. Zhou, and D. Chen, "Convolutional Neural Networks for Computer-Aided Detection or Diagnosis in Medical Image Analysis: An Overview," *Mathematical Biosciences and Engineering*, vol. 16, no. 6, pp. 6536–6561, 2019. https://doi.org/10.3934/mbe.2019326.
[39] Hon, Marcia, and Naimul Mefraz Khan. "Towards Alzheimer's disease classification through transfer learning." In *2017 IEEE International Conference on Bioinformatics and Biomedicine (BIBM)*, pp. 1166–1169. IEEE, 2017.
[40] S. Kloppel, C. M. Stonnington, J. Barnes, F. Chen, C. Chu, C. D. Good, I. Mader, L. A. Mitchell, A. C. Patel, C. C. Roberts et al., "Accuracy of Dementia Diagnosisa Direct Comparison Between Radiologists and Acomputerized Method," *Brain*, vol. 131, no. 11, pp. 2969–2974, 2008.
[41] Howard Lee and Yi-Ping Phoebe Chen, "Image Based Computer Aided Diagnosis System for Cancer Detection," *Expert Systems with Applications*, vol. 42, no. 12, pp. 5356–5365, July 2015.
[42] M. E. Celebi, Y. A. Aslandogan and P. R. Bergstresser, "Unsupervised Border Detection of Skin Lesion Images," *International Conference on Information Technology: Coding and Computing (ITCC'05)*, vol. 2, pp. 123–128, 2005. http://doi.org/10.1109/ITCC.2005.283.
[43] A. Esteva, K. Chou, S. Yeung et al., "Deep Learning-enabled Medical Computer Vision," *NPJ Digital Medicine*, vol. 4, no. 5, 2021. https://doi.org/10.1038/s41746-020-00376-2.

Index

0-9
10-fold cross-validation, 122–123

A
AlexNet, 181, 182, 183, 185, 186, 187
area under the curve (AUC), 117, 125
artificial intelligence (AI), 1–4, 7–9, 12, 182
artificial neural networks (ANN), 3, 4, 6
autism spectrum disorder (ASD), 167, 169–172, 174–177
AWS system, 200

B
big data (BD), 68–76
body sensor network (BSN), 39

C
cardiovascular disease, 138, 139, 142, 144, 145, 147, 151
cloud computing, 83–86, 88–95, 203
cloud integration, 204
computational intelligence (CI), 1–8, 9, 11–13
computed tomography (CT), 73, 196
computer-aided diagnosis, 195
computer vision, 194, 196, 197
convolutional neural network (CNN), 117–118, 139, 141, 143, 145, 147, 149, 151, 182, 190, 191, 196
coronavirus, 117–120, 132
Covid-19, 116–118, 122, 124, 128, 131–133, 198, 207
CT scan, 116–122, 132

D
data pre-processing, 115, 140, 141, 142
decentralized data routing, 99, 103
decision support system, 157–158
decision tree, 139, 145, 146, 150
deep learning, 24, 35–37, 39, 116–118, 122
diabetic foot ulcer, 181, 183, 190, 191
diagnosis, 1, 7, 9, 12, 15, 194, 198
diagnostic report generation, 105

E
ECG acquisition, 99–101, 112
ECG data routing, 102, 109
ECG diagnostic, 99–101, 103, 105
e-healthcare, 82–87, 89–90, 92–95, 99
electromyogram (EMG), 22
electronic health records (EHR), 73, 83, 85, 95

F
F1 measure, 125, 128, 131–132
feature extraction, 116, 119
fuzzy rule bases (FRBs), 170

H
Hadoop Distributed File System (HDFS), 73
healthcare, 1, 2–15, 18–31, 33, 35–36, 39, 41–45, 47–55, 58–60, 62–64, 68–76, 153, 156–158, 160
health informatics, 116–118, 121, 132
hyperparameter tuning, 118

I
image embedding, 116, 120–121
Inception v3, 116, 119, 122, 125–126, 128–129, 132
information and communication technologies (ICT), 23

Internet of Medical Things (IoMT), 23
Internet of Things (IoT), 18–23, 27–35, 41–45, 47–51, 53–56, 58–64, 84–88

K
knowledge transfer, 116

L
logistic regression, 139, 145, 150

M
machine learning, 2, 9, 18, 20, 22–23, 28–29, 35–37, 41, 48–51, 58–60, 64, 115, 117–118, 122, 132, 138, 139, 140, 142, 151, 152, 153–158, 160
magnetic resonance imaging (MRI), 73
medical data, 6–7, 14
medical imaging, 35, 38, 116–118, 131–133
multidimensional scaling (MDS) plot, 128–132
multi-layer perception (MLP), 20, 122

O
Omicron, 116
Orange toolkit, 118, 122, 124

P
painters, 119, 122, 125, 128–129, 131–132
pandemic, 115–116
pill descriptor vector, 198
prediction, 5, 8, 14
principal component analysis (PCA), 139

R
Raspberry Pi, 54, 56–57
remote assessment, 99, 112
remote monitoring, 50–51, 61
report routing, 99–100
ResNet, 180, 181, 182, 183, 184, 185, 186, 187, 188, 189
ROC curve, 126, 128, 133
routing algorithm at node, 99, 103

S
secure vector machine (SVM), 117, 121–122, 139
security, 60–61, 64
sensors, 50–51, 53–54, 56, 58–59, 61–62, 64
signal processing, 115
SqueezeNet, 118–119, 122, 125–126, 128–129, 131–132
supervised learning, 115, 118
surgery, 196

T
telemedicine, 61–62, 99–100, 113
training set, 116–122, 132
treatment, 197, 199

V
VGG-16, 116, 118, 122, 125, 127–128, 130, 132
VGG-19, 116, 122, 124–125, 127–128, 130, 132–133

W
wireless body sensor networks (WBSN), 22

X
X-ray, 116–122, 132

Printed in the United States
by Baker & Taylor Publisher Services

Printed in the United States
by Baker & Taylor Publisher Services